Common Inessential Discriminant Divisors

Scenes from the Early History of Algebraic Number Theory

HISTORY OF MATHEMATICS SOURCES VOL 47

Common Inessential Discriminant Divisors

Scenes from the Early History of Algebraic Number Theory

Fernando Q. Gouvêa
Jonathan Webster

2020 *Mathematics Subject Classification.* Primary 01A55;
Secondary 01A75, 11-03, 11R04.

Cover photo of Kurt Hensel courtesy of unknown photographer, via Archive of Philipps University Marburg. Cover photo of Richard Dedekind courtesy of ETH Library, public domain, via Wikimedia Commons. Frontispiece of Kurt Hensel article courtesy of Goettingen State- and University Library.

Library of Congress Cataloging-in-Publication Data

Names: Gouvêa, Fernando Q. (Fernando Quadros) author | Webster, Jonathan, 1978- author
Title: Common inessential discriminant divisors / Fernando Q. Gouvêa, Jonathan Webster.
Description: Providence, Rhode Island : American Mathematical Society, [2025] | Series: History of mathematics, 0899-2428 ; volume 47 | Includes bibliographical references and index.
Identifiers: LCCN 2025029615 | ISBN 9781470475246 paperback | ISBN 9781470481810 v. 47 ebook
Subjects: LCSH: Dedekind, Richard, 1831-1916 | Hensel, Kurt, 1861-1941 | Kronecker, Leopold, 1823-1891 | Algebraic number theory–History | AMS: History and biography – History of mathematics and mathematicians – 19th century | History and biography – History of mathematics and mathematicians – Collected or selected works; reprintings or translations of classics | Number theory – Historical | Number theory – Algebraic number theory: global fields – Algebraic numbers; rings of algebraic integers
Classification: LCC QA247 .G687 2025
LC record available at https://lccn.loc.gov/2025029615
History of Mathematics ISSN: 0899-2428 (print); 2472-4785 (online)
DOI: https://doi.org/10.1090/hmath/47

∞ The paper used in this book is acid-free and falls within the guidelines
established to ensure permanence and durability.
Visit the AMS home page at `https://www.ams.org/`

10 9 8 7 6 5 4 3 2 1 30 29 28 27 26 25

Contents

Preface

This is a book about a technical difficulty and an invitation to read the texts in which it was described, studied, and understood. These texts are from the first decades of what eventually became known as "algebraic number theory," though that name was not yet in use. Indeed, these papers helped create the subject.

In everyday life, technical difficulties tend to be annoying but unimportant. In mathematics, however, they are often the cause of important innovations. Once those innovations are made, the difficulty then provides a first test of the new methods. That was the historical role of common inessential discriminant divisors. Mysterious as they seemed at first, Richard Dedekind and Kurt Hensel managed to clarify when and why they occurred. This book is about how that happened.

Ernst Eduard Kummer developed a theory of "complex numbers built from roots of unity" around the middle of the 19th century. Leopold Kronecker and Richard Dedekind both found ways to generalize Kummer's theory. Dedekind published his version in 1871, but Kronecker held back until 1882.

In both cases, the resulting theory was surprisingly complicated. Dedekind's version involved several entirely new concepts: fields of algebraic numbers, algebraic integers, modules, and ideals. The theory Kronecker eventually published was far more general, treating algebraic numbers as a special case of "algebraic quantities," which included both algebraic numbers and algebraic functions. His method relied on polynomials in several variables.

As we will see, there was no great demand for a general theory at the time, so it was natural to wonder whether such complicated and abstract theories were really necessary. Both Dedekind and Kronecker pointed to common inessential discriminant divisors as evidence that the complication was inevitable.

It was Dedekind who first mentioned the issue in print, in 1871, in a notice **[20]** intended to call the attention of other mathematicians to the new results contained in Supplement X of the second edition of Lejeune Dirichlet's lectures in number theory. A few years later, having heard of E. I. Zolotarev's attempt **[143]** to create a theory of algebraic numbers based on "higher congruences," Dedekind published a crucial article **[23]** discussing index divisors and, in particular, giving proofs of the results he had announced in 1871. His goals seem to have been, first, to show that Zolotarev's approach could not work, second, to show that he understood what the problem was, and finally, to demonstrate that his ideal theory could give a sort of solution. We include annotated translations of both Dedekind texts, from 1871 and 1878, below.

Kronecker is a minor character in our story, but he does get a word in edgewise. He briefly discussed the problem in the last section of his *Grundzüge* **[97]** of 1882. In that section he also claims that the existence of common inessential disciminant divisors justifies the need for a new approach, in his case using polynomials in several variables. We include a translation of the relevant a portion of this section of Kronecker's *Grundzüge* below.

Kronecker assigned the problem of common inessential discriminant divisors to his student Kurt Hensel. One might guess that the motivation was similar to Dedekind's: given that the existence of common inessential discriminant divisors was one of the problems motivating his approach, it was now desirable to show that the new theory could describe why and when the problem happened. It seems likely, as well, that Hensel was simply more interested in algebraic numbers than his advisor.

Hensel's doctoral thesis (1884) dealt with the problem, but, as published, does not contain full proofs of his results. Hensel's further work on the problem formed a crucial part of his *Habilitation* in 1886. While his results were probably obtained in the 1880s, he only published them much later, after Kronecker's death. The key paper, whose contents seem not to have been completely described in the historical literature, is also given an annotated translation below. In it we see a restatement of the contention that the existence of common inessential discriminant divisors justifies the complications attendant on creating a theory of algebraic integers. Hensel re-obtains most of Dedekind's results from 1878 and goes further, obtaining a workable criterion for the existence of common inessential discriminant divisors.

Together, the texts we translate and discuss offer a snapshot of the mathematical activity of two key early contributots to the theory of algebraic numbers. They highlight the need to justify the new theories and the methods used by the two schools. Dedekind's *Anzeige* offers a good introduction to his methods, which he then put to work in 1878. Hensel's paper, on the other hand, used mostly Kroneckerian methods. Our annotated translation of Hensel's paper concludes the book.

We have attempted to produce translations that are mathematically precise, but have felt free to break up long Germanic sentences and paragraphs. In order to make it possible to read each individual paper on its own, we have allowed some repetitiveness into our annotations.

Richard Dedekind has been much studied, particularly by those interested in his philosophy of mathematics. The genesis of Dedekind's theory of ideals has also been extensively studied: see, for example, **[73]**, **[43]**, **[44]**, and **[2]**. These authors, however, point to other difficulties, particularly the need to define the notion of "algebraic integer" and to pin down the field discriminant. Dedekind certainly did both things, but he never described them as key obstacles on the way to ideal theory. Instead, he pointed to the existence of common inessential discriminant divisors.

Leopold Kronecker's theory of algebraic quantities has received less attention, both because he is more difficult to read and because Dedekind's methods won out in the end. Kronecker was by far the broader and more productive mathematician; his work encompasses many other topics. Kronecker's philosophy of mathematics, in particular, has received much attention. H. M. Edwards has focused on Kronecker's theory of algebraic quantities and in particular how it applies to algebraic numbers. Edwards highlights the philosophical aspects of Kronecker's choices (see, for example, **[41]**, **[47]** and **[46]**). Erwan Penchèvre has given a modern interpretation of Kronecker's general theory in **[116]**, emphasizing its connection to elimination theory and algebraic geometry. It is not our intention to go into the details and evolution of either approach except as it connects to the question on which we focus.

Kurt Hensel has received far less attention from historians. Peter Ullrich has written about Hensel's creation of the p-adic numbers in **[135]**, **[134]** and **[133]**. Birgit Petri's thesis **[117]** considers all of Hensel's early work with a view to charting his "way to the p-adic numbers." Petri includes a good summary account of Hensel's life.

The problem of common inessential discriminant divisors was fundamental in Kurt Hensel's early career and provided one of his motivations for introducing the p-adic numbers at the end of the 1890s. That story will be told in a forthcoming book by the first author.

Our annotations amount to a detailed mathematical reading of published sources. The published papers, however, also include indications of process; they allow us to trace, to some extent, the mathematical activity that led to their writing.[1]

For mathematicians interested in algebraic number theory we hope our annotated translations can give a glimpse of what their discipline looked like in this period. Hensel's paper, in particular, displays how Kroneckerian methods were used. These seem very foreign to us after the Noether-led triumph of Dedekind's approach.

Historians of mathematics may want to note how important it was to justify the introduction of new methods and new ideas. The differences between Dedekind and Hensel also highlight how much difficulty many mathematicians had with Dedekind's "simple" and "conceptual" methods. In particular, we note that Hensel (and perhaps also Kronecker) seem to have had real difficulty reading and understanding Dedekind, to the extent of giving new proofs of Dedekind's results. Throughout we suggest what amount to historical conjectures; perhaps historians will find it interesting to investigate them further.

[1]See **[50**, p. 27] for texts as traces of mathematical activity.

Acknowledgments

This book could not have been written without the internet and modern technology. Almost all of the papers we needed could be found online, and so we are grateful for the many repositories of historical mathematics journals. We depended heavily on TEX and LATEX, on SageMath **[123]**, on the *L-Functions and Modular Forms Database* **[109]** and the *Online Encyclopedia of Integer Sequeces* **[115]**, on translation software and online dictionaries, and much else. All of these things were created by dedicated people; we are in their debt.

Tim Molnar, then a student at Bates College, did some draft translations. Karl Stelzner, then at Butler College, did some as well. While our final translations preserve very little of their work, they certainly helped us get started, and we thank them. Many German-speaking friends helped us as well, including Renate Scheidler, Ina Mette, and Otto Bretscher.

Ina Mette was always a supportive editor. The anonymous referees helped us correct many mistakes, and the editorial board for the History of Mathematics series did an adroit job of handling a submission from one of its members. We thank you all.

Fernando would like to thank Jonathan for calling his attention to CIDDs and for keeping him focused as the writing progressed. He also thanks the Colby library's inter-library loan department. Mari was unfailingly supportive, especially in moments of discouragement.

Jonathan would like to thank Fernando for doing the final write-up and handling most of the interaction with the AMS.

We both had a great time working together to understand what Dedekind and Hensel were saying, creating numerical examples, and pondering what we could learn from all this, both about number theory and about the history of mathematics.

Finally, thank you, our readers. We hope you will enjoy the book.

CHAPTER 1

Setting the Stage

The story we want to tell begins in 1871 with the publication of Dedekind's Supplement X **[21]**, but every story arises from the past and every new idea has a pre-existing context. The goal of this chapter is to paint a picture of the context in which the work of Dedekind, Kronecker, and Hensel arose. To do this, we rely heavily on the work of historians of mathematics.

Modern mathematicians describe the subject we are interested in as "algebraic number theory," but this name was not in use in the 19th century. Neither was there really a coherent "subject": each of the historical actors had slightly different goals and problems in mind. The most common term used to talk about what we now call "algebraic integers" (an expression that seems to have been introduced by Dedekind) was simply "complex numbers," sometimes with a description of the specific kind of complex numbers under consideration.

Dedekind introduced the term "field" for certain subsets of the field of all complex numbers. Kronecker had an entirely different set of terms (we discuss them in Chapter 5) reflecting his different point of view. The term "algebraic number field" is due to Hilbert in his 1897 *Zahlbericht* **[87]**. Hilbert's influence seems to have established algebraic number theory as a mathematical subject, providing both a name and a set of goals to the discipline in the 20th century.

Since our goal is to provide background for reading the texts that follow, our focus is limited. We do not attempt a full account of number theory in this period, privileging instead those authors and texts that our historical actors would have read and studied. Even so, we have left out many details.

1.1. Stories and History

As in most intellectual communities, mathematicians tell stories about the past of their discipline. While these stories must bear some resemblance to what actually happened in the past, their primary goals are not historical. Research mathematicians tell stories to explain how their theorems fit into the flow of mathematical thought. Teachers of mathematics tell stories to create interest in their students. Doctoral advisors pass them on to budding mathematicians as a way of directing their research interests. The function of these stories is largely social: why you should read my paper, why my theorem is important, what the main questions are.

The social function of the narratives mathematicians tell and retell means that some of the messiness of real history must be avoided. The focus is inevitably modern. Concepts and ideas are presented in terms of where they eventually led, even if those later developments would have been unintelligible at the time. Avenues of research that did not lead anywhere need not be discussed. The narrative is teleological, often giving the impression that the historical actors knew or intuited where mathematics would eventually end up.

The story of how mathematicians came to be interested in algebraic numbers is no exception. There is a standard narrative of how number theory emerged as a separate subject early in the nineteenth century with the publication of Gauss's *Disquisitiones Arithmeticae*, developed over the century, and eventually arrived at the elaborate theory in Hilbert's *Zahlbericht*. It is a thoroughly German-centered story.

It is the job of historians of mathematics to query, correct, and complicate the story. A fundamental source for the modern historical account of number theory in the 19th century (especially in Germany) is *The Shaping of Arithmetic after C. F. Gauss's Disquisitiones Arithmeticae* [**66**], edited by C. Goldstein, N. Schappacher, and J. Schwermer. The opening chapter [**64**] provides an excellent overview of what mathematicians working on what was to become number theory found interesting in the first half of the century, while the second chapter [**65**] continues the story as far as Hilbert's famous summing-up. The rest of the book rounds out the story by looking at various specific topics and events. The account we give in this chapter is therefore heavily dependent on [**66**].

Catherine Goldstein and Norbert Schappacher summarize the "standard story" like this:

> Gauss's book (hereafter, we shall often use the abbreviation "the D.A." to designate it) is now seen as having created number theory as a systematic discipline in its own right, with the book, as well as the new discipline, represented as a landmark of German culture. Moreover, a standard history of the book has been elaborated. It stresses the impenetrability of the D.A. at the time of its appearance and integrates it into a sweeping narrative, setting out a continuous unfolding of the book's content, from Johann Peter Lejeune Dirichlet and Carl Gustav Jacobi on.
>
> In this history modern algebraic number theory appears as the natural outgrowth of the discipline founded by the *Disquisitiones Arithmeticae*. Historical studies have accordingly focused on the emergence of this branch of number theory, in particular on the works of Dirichlet, Ernst Eduard Kummer, Richard Dedekind, Leopold Kronecker, and on the specific thread linking the D.A. to the masterpiece of algebraic number theory, David Hilbert's *Zahlbericht* of 1897. [**64**, p. 4]

As expected, the historical investigations reported in [**66**] pretty much demolish this story. In particular, the authors show how the reception of the *Disquisitiones* was far more complex, how number theory in mid-century had far different concerns than the discipline Hilbert constructed, and indeed how there was very little demand for a theory of general "complex numbers" when it was created.

Goldstein and Schappacher describe their methodology thus:

> ...instead of going backwards, seeking in the *Disquisitiones Arithmeticae* hints and origins of more recent priorities, we will proceed forwards, following Gauss's text through time with the objective of surveying and periodizing its manifold effects. [**64**, p. 5]

This chapter, on the other hand, seeks to describe the background of the work done by Dedekind and Hensel between 1870 and 1895. For that purpose, we do look for "hints and origins," not of modern concerns, but of the problems that Dedekind and Hensel studied. We have tried, however, to tell the story in a way that respects what historians now know.

Moritz Epple, early in his magisterial account of the history of knot theory, points out that "No even moderately complex episode of mathematical activity admits only a single apt description."[1] We hope our description is sufficient to give readers a sense of the context in which Dedekind and Hensel wrote.

1.2. Signposts

As the nineteenth century began, two books summarized the past and delineated the future of what was eventually to become a new mathematical discipline. Adrien-Marie Legendre's *Essai sur la Théorie des Nombres* (1798, 2nd. ed. 1808, 3rd. ed. 1830) [**105**] gave it one of its names, the Theory of Numbers. Carl Friedrich Gauss's *Disquisitiones Arithmeticae* (1801, reprinted 1863 in [**57**]) established the other name, the Higher Arithmetic. Legendre's "essay" was more conservative, reflecting and extending the achievements of Euler, very much focused on Diophantine equations and their solutions. Gauss's "investigations" were more ambitious, innovative, and difficult to read. They were also more limited: Gauss specifically excluded Diophantine equations from his work.

Both books pointed the way to many of the central concerns of the future. Three elements turned out to be especially important. Legendre stated and tried to prove the quadratic reciprocity law, but it was Gauss who first gave a correct proof (indeed, several of them). Gauss went on to study biquadratic reciprocity, leading others to search for other generalizations. Legendre and Gauss both studied binary quadratic forms, building on the work of Euler and Lagrange. This put the arithmetical theory of quadratic forms on center stage. Gauss's final chapter on cyclotomy, which he described as not part of the main subject but rather an application, was the third fundamental topic.

Two very different texts can serve as signposts for the situation in mid-century. The first edition of Dedekind's write-up of Dirichlet's lectures, *Vorlesungen über Zahlentheorie*,[2] [**34**] was published in 1863. Dirichlet's lectures provide a more accessible account of parts of Gauss's *Disquisitiones*, focusing especially on the theory of binary quadratic forms. Unlike Legendre's and Gauss's books, Dirichlet-Dedekind was a textbook, focused on standard material. Newer research results (mostly due to Dirichlet himself) appeared only in the nine "supplements," written by Dedekind but based on Dirichlet's papers. Dirichlet's *Vorlesungen* therefore offers us one view of what was considered "basic knowledge" of number theory.

The other mid-century work that deserves some attention was very different. H. J. S. Smith's *Report on the Theory of Numbers* was published in six parts as a series of reports to the British Association, between 1859 and 1865. (See the much later reprint in one volume [**129**].) It is focused on research and provides us with a contemporary view of recent results in number theory. Smith's report opens with a summary of the basic theory of congruences and elementary number theory as given in the Legendre's *Essai* and Gauss's *Disquisitiones*, then goes on to trace what he saw as the main developments during the first half of the 19th

[1] Keine auch nur einigermaßen complexe Episode mathematischen Handelns läßt nur eine einzige treffende Beschreibung zu. [**50**, p. 23]

[2] See [**61**] for details on the book and its history.

century. These consist of two main topics: reciprocity laws and the theory of forms, especially binary quadratic forms. Cyclotomy, infinite series, and elliptic functions appear mostly as tools aimed at understanding the two main topics. Writing just before Dedekind first introduced his ideal theory, Smith gives one view of what was important at the time.

Legendre died in 1833[3], but Gauss remained active until the 1850s. By 1830, Abel had appeared briefly on the scene but was gone. The major players over the next two decades were Jacobi, Dirichlet, Kummer, Hermite, and Eisenstein, with Galois not really influential until the late 1840s. While their interests varied, much of their work fits together into a coherent mathematical research program that Goldstein and Schappacher call "arithmetic algebraic analysis."

At the center of this group of mathematicians stood Gustav Lejeune Dirichlet.[4] All the main players had some direct connection to him. Dirichlet lived in Paris for a while; there he and Legendre proved the degree five case of Fermat's Last Theorem in a sort of friendly competition/collaboration (see [**111**, ch. 3] for the story). Legend[5] has it that Dirichlet carried a copy of Gauss's *Disquisitiones* everywhere he went, and certainly Gauss was instrumental in getting him to return to Germany (to be precise, to Prussia). Dirichlet moved to Berlin in 1828; from there he influenced and interacted with many others. Jacobi was a personal friend and correspondent. Hermite, late in life, described himself as a "disciple" of Gauss, Jacobi, and Dirichlet.[6] Eisenstein studied with Dirichlet in the 1840s. Kummer's first wife was a cousin of Rebecca Dirichlet.[7] Dirichlet had a role in bringing Kummer to Berlin (as his own replacement). Kronecker studied with Kummer early on, then moved to Berlin, completing his doctorate in Berlin in 1845 under Dirichlet. In 1855, Dirichlet went to Göttingen (to replace Gauss after his death) and ended up having a huge influence on the young Richard Dedekind and Bernhard Riemann. Dedekind's thesis was completed in Göttingen under Gauss's direction in 1852, but he claimed to have learned most of his mathematics from Dirichlet.[8]

Dirichlet's mathematical ideas were similarly important. Today he is perhaps best known for being the first to use complex analysis and Fourier analysis to prove results in number theory, with striking success. These methods were the key to his class number formula, a major result on binary quadratic forms, and to the proof that if $\gcd(a, b) = 1$ there exist infinitely many prime numbers of the form $a + bx$. (See [**111**, ch. 9] for summaries of some of these papers.) But Dirichlet worked on many other things, obtaining in particular the theorem describing "complex units" (invertible elements in a ring of algebraic integers).

What did mathematicians interested in "the theory of numbers" study? There was a wide range of topics that interested them, as shown in [**64**]. Many of those topics arose from Gauss's *Disquisitiones*, but they were pursued separately and with

---

[3]See the table at the end of this chapter (page 19) for full names and dates.

[4]See [**111**] for a recent mathematical biography. We mostly follow Merzbach's account.

[5]According to [**64**, p. 45] the story was spread by Kummer.

[6]"J'ai toujours été et je serai jusqu'à mon dernier jour le disciple de vos grands géomètres Gauss, Jacobi, Dirichlet." Letter to Eugen Jahnke Quoted by Catherine Goldstein in [**62**]; see also [**63**]. Kronecker, Borchardt, and Lipschitz are described as "compagnons" and "amis," however.

[7]Rebecca Dirichlet was the younger sister of Fanny Hensel and Felix Mendelssohn, who had already made their marks in the world of music. Through her, Dirichlet would have been in contact with the wider cultural world, particularly in Berlin.

[8]See, for example, the discussion in [**44**].

many different goals in mind. We highlight only those topics connected more or less directly with algebraic numbers; readers should consult historical works such as **[66]** for a broader perspective.

1.3. What Gauss Left Undone: Higher Congruences

All attentive readers of Gauss's *Disquisitiones* would have noticed that the eighth chapter was missing. At several points in the book Gauss mentions it and tells us that it would deal with "higher congruences," meaning (at least) congruences $f(x) \equiv 0 \pmod{p}$ where p is a prime number and $f(x)$ is a polynomial of degree bigger than two. That chapter would have nicely closed the circle,[9] since the first chapter had introduced congruences and dealt with the case when $f(x)$ is of degree one, and subsequent chapters had, by giving a proof of the law of quadratic reciprocity, settled the problem when $f(x)$ has degree two. But the eighth chapter is not there.

It is unclear why this material ended up not being included.[10] After Gauss's death an unfinished draft of the eighth chapter was found; when the second volume of Gauss's collected works was published in 1863, this was included.[11]

By that time, several authors had done their own research into higher congruences. Most important for our purposes is Dedekind's "Outline of a Theory of Higher Congruences with Respect to a Real Prime Modulus" (Abriß einer Theorie der höheren Kongruenzen in bezug auf einen reellen Primzahl-Modulus **[18]**), published in *Crelle's Journal* in 1857, before Dedekind had seen Gauss's draft. This was intended to summarize, organize, and go beyond the work of Galois, Serret, and Schönemann.[12]

Dealing with polynomial congruences modulo p one immediately sees that they do not always have roots in $\mathbb{F}_p$. Galois in **[56]** was willing to simply postulate "imaginary" roots, but Gauss (and the German mathematicians who followed him) explicitly refused to do so (see **[54**, p. 180] for a representative quote). Dedekind had clearly read Galois, but he followed Gauss: rather than working with "imaginaries," he introduced congruences with respect to a "double modulus" (p, P), where p is a prime number and P is a (usually irreducible) polynomial in one variable with integer coefficients:

$$f(x) \equiv g(x) \quad \text{modd.}\,(p, P)$$

if $f(x) - g(x) = P(x)h(x) + pk(x)$, where $h(x)$ and $k(x)$ are polynomials with integer coefficients. In modern language and in general, Gauss's theory focuses on the ring $\mathbb{F}_p[x]$ of polynomials over the field $\mathbb{F}_p$ while Dedekind also goes on to study the finite fields[13] $\mathbb{F}_p[x]/(P)$. Two results in this paper are worth noting.

---

[9]See **[64**, 1.4] for a discussion of the structure of the *Disquisitiones* and how chapter eight would have played a concluding role.

[10]In his introduction, Gauss just says he decided not to include it because it would have made the book too long. This may be related to the fact that the chapter on quadratic forms seems to have "roughly doubled in length" **[64**, p. 5] during the printing process.

[11]See **[54]** for a detailed analysis of the contents of the draft chapter 8.

[12]Dedekind just mentions their names, but the editors of his collected works provide references to **[56]**, **[128**, pp. 343–370], **[125]**, and **[126]**. But see also **[75]**, where Hensel gives his own take.

[13]He did not, however, call them "finite fields" or "fields" at all. Doing so would have required thinking of congruence classes as objects in themselves, which was simply not done at the time. In Dedekind the term "finite field" refers to subfields of the complex numbers that are finite-dimensional over $\mathbb{Q}$.

First, Dedekind showed that if the polynomial P had degree n and was irreducible modulo p, the congruence

$$x^{p^n} \equiv x \mod p, P$$

was true for every x. Second, he used this result to find a formula giving the number, up to congruence, of polynomials of degree n that are irreducible modulo p. This would eventually play a role in the story of common inessential discriminant divisors.

Galois thus worked with some kind of finite fields,[14] and Serret discussed them in his book [**128**]. Except for Selling, all of the German mathematicians discussed in this book worked instead with congruences with respect to a double modulus (p, P). As a result, when studying algebraic numbers, Dedekind, Kronecker, and Hensel had to come up with ways to relate algebraic integers modulo a prime divisor/ideal to congruences modulo (p, P) for some polynomial P. (See Chapter 4 for examples in Dedekind's case.)

1.4. Cyclotomy and Elliptic Functions

The last chapter of Gauss's *Disquisitiones* dealt with the arithmetic properties of roots of unity and their application to the classical problem of constructing regular polygons. The division of the circle ("cyclotomy") led to polynomials (the "cyclotomic polynomials") whose algebraic properties are especially simple. Using sums of roots of unity, Gauss also introduced "periods" that we now understand as canonical generators of the subfields of the cyclotomic field $\mathbb{Q}(\zeta)$ generated by an n-th root of unity.

Gauss famously used his cyclotomic theory to establish which polygons could be constructed with straightedge and compass. More significant for the future of number theory, however, was his application of roots of unity to give another proof of quadratic reciprocity.

Cyclotomy quickly became a central topic in number theory. In 1836, Jacobi[15] described "the theory of the solution of pure equations" (by which he meant equations of the form $x^n - a = 0$, and so cyclotomy and reciprocity) as one of the two major branches of the subject. (The other was the theory of quadratic forms.) Gauss, Eisenstein, and Jacobi all explored the sums that could be constructed with roots of unity and their connections to other topics in number theory. These included Gauss's periods and what we now call "Gauss sums" and "Jacobi sums." A modern summary of the results is given by André Weil in [**139**].

This leads us to the second topic that Gauss promised to write about (but never did). From the point of view of the time, cyclotomy amounted to a surprising link between the transcendental functions sine and cosine (which parametrize the unit circle) and arithmetic. Early in chapter 7 of the *Disquisitiones* ([**59**, Art. 335, p. 407]), Gauss comments[16] that the ideas could easily be extended to other transcendental functions, for example "those which depend on the integral $\int \frac{1}{\sqrt{(1-x^4)}}\,dx$." This well-known elliptic integral goes back to the work of L. Euler and G. Fagnano; it arises from computing the arc length of the lemniscate (see, for example, [**3**] and [**120**]). Gauss then adds "Since, however, we are preparing a

---

[14]Not, of course, using that name.

[15]Cited in [**64**, p. 28] from handwritten lecture notes from Jacobi's *Theorie der Zahlen* course, but now see [**95**, p. 8].

[16]Goldstein and Schappacher call attention to Gauss's comment in [**64**, p. 32].

large work on those transcendental functions and since we will treat congruences at length in the continuation of these *Disquisitiones*, we have decided to consider only circular functions here."[17]

The "large work" was never published, but many authors, beginning with Abel and Jacobi, picked up on the hint.[18] Gauss's imprimatur helped make elliptic functions and their division polynomials a central topic for nineteenth century mathematics. Legendre had published a three-volume treatise on elliptic integrals, and his work was transformed and developed further by Abel, Jacobi, Hermite, Eisenstein, Kronecker, and many others.

There was particular interest in elliptic functions (like those associated to the integral Gauss mentioned) that have the property that became known as "complex multiplication." These turned out to have deep connections to the theory of binary quadratic forms of negative discriminant.[19]

The plethora of identities satisfied by elliptic and theta functions fascinated many number theorists. Many of these identities turn out to contain arithmetical information. Both the theory of reciprocity laws and the theory of quadratic forms were deeply impacted by these methods.

1.5. Binary Quadratic Forms and Generalizations

The longest chapter of Gauss's *Disquisitiones* dealt with the theory of binary quadratic forms (with just a hint of ternary quadratic forms as well).[20] Building on prior work by Lagrange and Legendre, Gauss created a large and impressive theory. Much of what we now think of as the theory of quadratic fields is already there in implicit form. It is phrased, however, entirely in the language of polynomials with integer coefficients. Gauss's theory turned out to be very rich and caught the attention of many mathematicians. In particular, it was natural to attempt to generalize it to forms in n variables of degree m, and perhaps also to forms with non-integer coefficients. Smith dedicates the greater part of his *Report* to this subject; see [**129**, Parts III to VI]. For a modern mathematical account of the work of Lagrange, Legendre, and Gauss on the subject see [**17**, § 2 and 3].

Gauss's treatment of binary quadratic forms was very difficult, leading to many attempts to give a simpler account. (As we mentioned above, this was one of the goals of Dirichlet's *Vorlesungen* [**34**]. A modern attempt to do the same is [**53**]; a more extensive account can be found in [**17**].) In many modern treatments the theory developed by Gauss gets translated into the language of quadratic number

---

[17] Nanque non solum ad functiones circulares, sed pari successu ad multas alias functiones transcendentes applicari possunt, *e. g.* ad eas, quae ab integrali $\int \frac{1}{\sqrt{(1-x^4)}}\,dx$ pendent; ... sed quoniam de illis functionibus transcendentibus amplum opus peculiare paramus, de congruentiis autem in continuatione disquisitionum arithmeticarum copiose tractatibur, hoc loco solars functiones circulares considerare visum est.

[18] A full history of the theory of elliptic, modular, and theta functions remains to be written, though of course there are many accounts of aspects of the theory. A good overall summary is [**90**]. See also [**91**, ch. IX], [**93**], [**92**], [**122**], [**136**], [**138**].

[19] A modern exposition of large parts of the theory can be found in [**17**].

[20] We use "form" to mean a homogeneous polynomial of degree m in n variables. Binary quadratic forms are the case $m = n = 2$; ternary quadratic forms have $m = 2$, $n = 3$. Most mathematicians in the 19th century used the word in this sense.

fields;[21] Gauss and his successors avoided doing that, though some of them (notably Kummer; see [**48**, 5.1]) do seem to be aware of the possibility.

A key question was "representation" by quadratic forms; for example, what prime numbers p could be represented as $x^2 + 7y^2$ with $x, y \in \mathbb{Z}$? If such a representation exists, can we find it? Is there more than one? If so, can we find them all? Are there explicit formulas? As the theory developed, interest (particularly Gauss's interest) shifted from the numbers represented by the quadratic forms to the forms themselves.

If an invertible (over $\mathbb{Z}$) change of variables transforms one form into another, the two forms will represent the same numbers, so it is natural to introduce some notion of equivalence.[22] Equivalence preserves a key invariant of the form, a number now called the *discriminant*. As Legendre already knew, for a fixed discriminant one can find a finite number of "reduced forms" with the property that any other form is equivalent to one of them.

Gauss worked with proper equivalence of binary quadratic forms and studied the resulting equivalence classes.[23] Since for a fixed discriminant there are finitely many classes, determining the exact number (called the "class number" and traditionally denoted by h) became one of the key problems.

A notion of composition of forms had been introduced by Lagrange and Legendre to generalize such identities as

$$(x^2 + ny^2)(u^2 + nv^2) = (xu - nyv)^2 + n(xv + yu)^2.$$

Given two binary quadratic forms $f(x, y)$ and $g(x, y)$ with the same discriminant, their composition should be a binary quadratic form $F(x, y)$ with the same discriminant such that

$$f(x, y)g(u, v) = F(a(x, y, u, v), b(x, y, u, v))$$

with a and b being bilinear functions of (x, y), (u, v). One result would be that if a number m is represented by $f(x, y)$ and a number n is represented by $g(x, y)$, then the number mn is represented by $F(x, y)$. Legendre showed that the composed form $F(x, y)$ always exists, but knew that it is usually not unique (not even unique up to equivalence).

Gauss introduced a restriction on the bilinear functions a and b to obtain a uniquely determined composition of (proper equivalence classes[24] of) binary quadratic forms. Gauss showed that the finitely many classes form a group under composition,[25] though of course he did not use that term.[26] The "principal class" was the identity in that group. The class number h, is the order of this finite group,

[21] Some modern authors, notably André Weil, argue that Gauss was aware of the equivalence between his theory and the arithmetic of quadratic fields.

[22] The term "equivalence" in this context is due to Gauss. Restricting to changes of variable with determinant one defines "proper equivalence,"

[23] He seems to have borrowed the word "class" from biology; forms also belonged to a "genus" and an "order."

[24] One must note, however, that Gauss never speaks of the composition of classes. He was willing to work with *forms* as primary objects, but he never took the step of thinking of *classes* as objects in their own right. Most of the mathematicians we discuss followed him on this point: forms were composed, not classes. They worked with congruences, not congruence classes. The modern impulse to immediately pass to the quotient simply did not exist in the 19th century.

[25] This is far from obvious. Gauss's proof that composition of classes is associative in [**59**, § 240] is famous for its difficulty.

[26] Not only was the notion of group still in the future, but also he would never had treated the equivalence classes as elements of some set.

so that composing a form with itself h times always gave a form from the principal class. Gauss stressed the similarity between equivalence classes of forms and the nonzero congruence classes modulo a prime number.[27]

In the case of positive definite binary quadratic forms, several of these questions turned out to be connected to elliptic and theta functions with complex multiplication. (Part IV of Smith's *Report* is almost entirely about these results; it is the goal of [**17**] to give a modern account of them.) The nature of the connections was initially rather obscure, so mathematicians tried to clarify and extend them.

Many authors attempted to generalize Gauss's theory of binary quadratic forms to homogeneous polynomials of degree m in n variables. Generalizing the notion of equivalence offered little difficulty, but a theory of composition of forms was not that easily found.[28] Some aspects of the theory quickly became more algebraic than arithmetical, leading to invariant theory and algebraic geometry.

A few authors considered forms of a special type that are closely related to what would become algebraic number theory. Let $\varphi(x)$ be an irreducible monic polynomial of degree n with integer coefficients, and let $\alpha_1, \alpha_2, \ldots, \alpha_n$ be its roots. Let $f(y) = x_1 + x_2 y + \cdots + x_n y^{n-1}$, where the x_i are thought of as indeterminates. Then the product

$$N(x_1, x_2, \ldots, x_n) = f(\alpha_1) f(\alpha_2) \cdots f(\alpha_n)$$

is a homogeneous form in n variables with integer coefficients. (We now call this a *norm form* for the number field $\mathbb{Q}(\alpha_1)$.) Not all forms of degree n in n variables have the property that they are the product of n linear forms of degree one with irrational coefficients, so such forms were called "decomposable." Studying decomposable forms was, then, an indirect way of studying algebraic integers, much as binary quadratic forms give an indirect way of studying quadratic fields. Both Hermite and Dirichlet considered such forms. In fact, Dirichlet's "unit theorem" was phrased in terms of finding solutions to the equation $N(x_1, x_2, \ldots, x_n) = 1$ ([**36**], [**38**]). In the introduction to [**86**], Hermite thinks of the theory of decomposable forms as "the same" as a general theory of complex numbers:

> These results are about complex numbers, considered in general, or rather to the theory of certain forms that are decomposable into linear factors...[29]

As we will see, Dedekind makes the same connection when he sets out to introduce his general theory of algebraic integers.[30]

1.6. Determinant and Discriminant

Since we have just mentioned the discriminant of a binary quadratic form and since we later will say much about discriminants of polynomials, it is good to briefly

[27]He seems to have been surprised to find that the class group is not always cyclic.

[28]Higher composition laws are still a topic of research.

[29]Ces résultats sont relatifs surtout aux nombres complexes, considérés en general, ou plutôt à la théorie de certaines formes décomposables en facteurs linéaires...

[30]In his review of Eisenstein's *Mathematische Werke*, Weil argues that "The example of Gauss' theory of binary quadratic forms had misled his immediate successors into believing that the key to algebraic number-fields was to be found in the study of forms of degree n in n variables over $\mathbb{Z}$, decomposable into n linear factors; this seems to have been Jacobi's and Dirichlet's view; perhaps it was Gauss' opinion." [**141**, vol. III, p. 400]

review the history of these notions. We do no more than note the main stepping-stones, pointing the reader to [**112**] for an exhaustive account.

Gauss's word for the discriminant of a binary quadratic form was "determinant." He introduced the term in Article 154 of the *Disquisitiones*, extending it to ternary quadratic forms in Article 267. In both cases Gauss simply wrote out a polynomial in the coefficients of the form, giving little explanation of its form.

In Gauss's second proof of the "Fundamental Theorem of Algebra" [**60**], published in 1815, he gave a definition of the "determinant" of a polynomial in one variable. He started from the generic polynomial with m roots[31]

$$(x-a)(x-b)(x-c)\cdots = x^m - \lambda' x^{m-1} + \lambda'' x^{m-2} - \lambda''' x^{m-3} + \text{etc.}$$

then considered the product

$$(a-b)(a-c)\cdots(b-a)(b-c)\cdots(c-a)(c-b)\cdots\text{etc.}$$

of all differences of the roots. This is a symmetric function of the roots, hence (by a theorem Gauss proved in the same paper) is polynomial in the coefficients $\lambda', \lambda'', \lambda'''$. Because he wanted to be able to define this quantity without assuming that a polynomial factors into terms of degree one, Gauss then replaced each λ by a new indeterminate ℓ (with the same number of primes) to get a polynomial in $\ell', \ell'', \ell''', \ldots$. This he defined to be the "determinant" of the polynomial

$$x^m - \ell' x^{m-1} + \ell'' x^{m-2} - \ell''' x^{m-3} + \ldots$$

He wrote out the resulting polynomials when $m = 2$ and $m = 3$, but said nothing more about how to find these polynomials explicitly.

Gauss went on to note that for polynomials f that factor into terms of degree one it is obvious that the discriminant is zero if and only if $\gcd(f, f') \neq 1$. He went on to give a proof of this fact *without* assuming that the polynomial factors.

This definition is essentially the same as the modern definition, except that rather than using $(a-b)(b-a)$ we now use $(a-b)^2$, so that Gauss's definition and ours differ by a sign. Gauss's definition applies only to monic polynomials, but it is easy to see how to generalize.

The "determinants" of binary forms are, of course, the same as the "determinant" of the polynomial we get by dehomogenizing. For example, the "determinant" of the quadratic form $ax^2 + bxy + cy^2$ is $b^2 - 4ac$, which is also the "determinant" of the polynomial $ax^2 + bx + c$ corresponding to it.[32]

Almost simultaneously with this paper of Gauss, Augustin-Louis Cauchy was generalizing the "determinant" in a different way. In [**14**], Cauchy defined the determinant of an $n \times n$ array of numbers, telling us that he was inspired by ideas of Vandermonde and Bézout in elimination theory and by Gauss's notion of the determinant of a form. Cauchy's determinant of an array is the same quantity we now call the determinant of an $n \times n$ matrix (but matrices came later). Determinants quickly became an important algebraic tool and were the subject of a lot of research work, as the size of Muir's [**112**] itself demonstrates.

One of the champions of the determinant was J. J. Sylvester, who systematized its connection to elimination theory and used it in his invariant theory work.

---

[31] We follow Gauss's notation in this paragraph.

[32] Gauss actually operated with binary quadratic forms with even middle coefficient, $ax^2 + 2bxy + cy^2$, so for forms his "determinant" is $b^2 - ac$, which is one fourth of the "determinant" of the associated polynomial.

Sylvester was a great giver of names, and his 1851 article **[130]**, he pointed out the inconvenience of having two different objects sharing the name "determinant." He suggested renaming the "determinant" of a form or of a polynomial the *discriminant* instead.

Thus we get the modern definition: if f is a monic polynomial of degree n with roots $r_1, r_2, \ldots, r_n$, we define

$$\operatorname{disc}(f) = \prod_{i<j}(r_i - r_j)^2.$$

(If $f(x)$ is not monic, multiply by the $(2n-2)$th power of the leading coefficient.) Notice that this is the square of the Vandermonde determinant whose columns are the powers of the roots. The discriminant is given by a polynomial in the coefficients of the polynomial f with integer coefficients. It can also be computed in several different ways, for example using the resultant of the polynomial f and its formal derivative f'.

The new name became standard in English and spread fairly quickly. Hermite was already using it (for quadratic forms) in 1854, attributing it to "English geometers" ("les Géomètres anglais"; **[83]**, **[85**, p.225**]**). But in 1857, we see Hermite glossing the word "discriminant" as "Gauss's determinant,"[33] showing that the term was not yet universally known. Richard Balzer, in his textbook *Theorie und Anwendung der Determinanten*, used "determinant" in the first three editions (1857, 1864, 1870) but switched to "discriminant" in the fourth edition of 1875, saying in a footnote that "in current usage the name "discriminant" is more common."[34]

Kronecker used the term "determinant" in **[99]**, an 1854 paper that Dedekind described as the "first hint" ("erste Spur," **[21**, p. 447**]**) that polynomial discriminants were important in number theory.[35] By 1881 in **[100]**, he had switched to "discriminant," which he also used in the *Grundzüge*. In Chapter 3 we will see how in 1871 Dedekind introduced still another kind of "discriminant."

1.7. Reciprocity, Quadratic and Higher

The quadratic reciprocity law was conjectured by Leonard Euler and stated in its modern form by Legendre. In his *Essai* **[105**, pp. 198ff**]**, Legendre attempted to give a proof, which he then "improved" in successive editions without ever really getting it right. The first proofs that convinced other mathematicians were those given by Gauss in the *Disquisitiones*. Gauss gave several proofs and added further proofs in other papers. A remarkable aspect of these proofs was the huge variety of techniques they seemed to involve: induction, counting, cyclotomy, the genus theory of quadratic forms, etc. The depth of the reciprocity law was suggested by the fact that it could be linked to many other topics.[36]

---

[33] He says "le discriminant (déterminant de Gauss)." See **[84]**, **[85**, p. 415**]**

[34] Bei dem jetzigen Sprachgebrauch ist der von Sylvester (Philos. Mag, 1851. II p. 406) gebildete Name "Discriminante" bezeichnender (Footnote on p. 123 of **[6]**).

[35] Kronecker's theorem was that the n-th cyclotomic polynomial was not only irreducible over $\mathbb{Q}$ but remained irreducible over $\mathbb{Q}(\alpha)$ if α is the root of an irreducible polynomial whose discriminant is prime to n.

[36] The best historically-based account of the theory of reciprocity laws is by Franz Lemmermeyer **[106]**; see also **[9]**. Reciprocity is the main focus of the first two parts of Smith's *Report* **[129]**.

It seemed natural to go on to ask about more general congruences $x^n \equiv a \pmod{p}$ and to investigate whether there were also degree n reciprocity laws. Gauss led the field when he published two memoirs on biquadratic reciprocity (1825, 1831). This set off a quest for more and more general power reciprocity laws.

It was in the context of biquadratic reciprocity laws that Gauss introduced what are now called the "Gaussian integers": the ring $\mathbb{Z}[i]$ consisting of complex numbers of the form $a + bi$ with $a, b \in \mathbb{Z}$.[37] The reasons for that move are clear.[38] In the quadratic theory one has the Legendre symbol $\left(\frac{a}{q}\right)$, taking values ± 1, that specifies whether or not a is a quadratic residue modulo a prime number q. Two properties play a crucial role. First, this symbol is multiplicative:

$$\left(\frac{ab}{q}\right) = \left(\frac{a}{q}\right)\left(\frac{b}{q}\right).$$

Second, there is a congruence relation (usually attributed to Euler):

$$a^{(q-1)/2} \equiv \left(\frac{a}{q}\right) \pmod{p}.$$

Gauss realized that in order to have a multiplicative biquadratic residue symbol it would need to have four, rather than two, states. To achieve this, he defined the symbol to take values *in the fourth roots of unity.* Then, in order to obtain a congruence analogous to Euler's, one must operate in $\mathbb{Z}[i]$. It quickly became clear that even the statement of a proper biquadratic reciprocity law depended on the arithmetic of "complex numbers" in $\mathbb{Z}[i]$. Gauss established the basic theorems about their arithmetic of (this kind of) complex numbers.[39] A short time later, Eisenstein followed his steps by studying cubic reciprocity, which depended on the arithmetic of "complex numbers depending on a cube root of unity."

The variety of the proofs of quadratic and higher reciprocity laws was striking. Jacobi and Eisenstein, for example, used both cyclotomic and elliptic variants of Gauss sums in their proofs. These connections served to emphasize the centrality of reciprocity.

1.8. Kummer's Cyclotomic Theory

It was the search for more general reciprocity laws that led Kummer to generalize Gauss's and Eisenstein's ideas by considering "complex numbers built from n-th roots of unity." For the same reasons discussed above, an n-th power reciprocity law required working in $\mathbb{Z}[\alpha]$, where α is a complex number such that $\alpha^n = 1$ but $\alpha^k \neq 1$ for $0 < k < n$. A short paper by Jacobi reflects this consensus, discussing "the prime complex prime numbers that are considered when studying residues of degree 5, 8, or 12."[40] The idea is that to study higher reciprocity laws requires first understanding the arithmetic of such complex numbers. Jacobi's paper offers little

[37] In the early 1800s these were usually referred to as "complex numbers," or, as more precision became necessary, "complex numbers built from the square root of -1."

[38] At least they seem clear to us. Jacobi, in **[94]** disagreed, saying that the strange idea of working with complex numbers must have arisen in the context of the theory of division of the lemniscate. See **[64**, p. 40**]**.

[39] Dirichlet seems to have been struck by the similar properties of $\mathbb{Z}$ and $\mathbb{Z}[i]$, writing several papers exploring $\mathbb{Z}[i]$-versions of standard number theory.

[40] Ueber die complexen Primzahlen, welche in der Theorie der Reste der 5ten, 8ten und 12ten Potenzen zu betrachten sind **[94]**. Goldstein and Schappacher highlight this paper in **[64]**.

more than hints, however; more details could have been found in the notes from his lectures on number theory.[41]

Around this time, Kummer set out to study the arithmetic of complex numbers built from n-th roots of unity, initially under the assumption that n is prime. He worked with numbers of the form

$$a_0 + a_1\alpha + a_2\alpha^2 + \cdots + a_{n-1}\alpha^{n-1},$$

where $a_1, a_2, \ldots, a_{n-1}$ are integers. One can add and multiply such numbers, and so one can study their factorization. As is well known, Kummer eventually realized[42] that there was a crucial problem: factorization into irreducible factors was in general not unique.[43] Kummer's description of the problem has become standard: he says that when the prime n is 23 there are irreducible numbers (i.e., numbers that cannot be factored non-trivially) that do not behave as primes should (i.e., they can divide a product of two numbers without dividing either factor). Such irreducible numbers do not have "the true nature of prime numbers" (... noch nicht die wahre Natur einer complexen Primzahl hat; see **[104]**.) An examination of the usual proof of unique factorization for the integers shows that this is the crucial ingredient, so Kummer's observation shows that if we try to factor in terms of irreducible numbers, factorization will not be unique.

To "save" the property of unique factorization, Kummer came up with a daring interpretation of the situation, invoking an analogy from chemistry: just as chemists could tell there were atoms and molecules without directly observing them, there must be cyclotomic "primes" that we cannot directly observe.[44] When, for example, one sees that

$$f(\alpha)g(\alpha) = h(\alpha)k(\alpha)$$

and the numbers $f(\alpha), g(\alpha), h(\alpha), k(\alpha)$ are all irreducible, then there must be unobservable factors that are getting recombined:

$$\begin{aligned} f(\alpha) &= \varphi\psi & \qquad h(\alpha) &= \varphi\epsilon \\ g(\alpha) &= \epsilon\eta & \qquad k(\alpha) &= \psi\eta \end{aligned}$$

One cannot simply postulate the existence of "ideal factors," however, so Kummer developed a way of testing whether a complex number $f(\alpha)$ was or was not divisible by an ideal prime $\mathfrak{p}$, and indeed showed that one can define precisely the maximal power n such that $\mathfrak{p}^n$ divides $f(\alpha)$ but $\mathfrak{p}^{n+1}$ does not. In effect, the divisibility test *defines* the ideal prime.[45] The definition is algorithmic: to check whether $f(\alpha)$ is divisible by $\mathfrak{p}$, one multiplies $f(\alpha)$ by a well-chosen number $h(\alpha)$ and checks whether the result is divisible by an integer prime p. (The number $h(\alpha)$ acts as a catalyst

---

[41] These have been published only recently, as **[95]**, but seem to have been available to Kummer. See **[107]**.

[42] Lemmermeyer argues in **[107]** that Dirichlet, Jacobi, and Eisenstein probably knew this before Kummer.

[43] The history of Kummer's investigations is well summarized in **[64]**. For an account of their content, one might consult Part II of Smith's *Report*, Articles 40–61 **[129**, pp. 94–137], Lemmermeyer's account in **[107]**, André Weil's introduction to **[102]**, the review **[110]**, and Chapters 4–6 of Edwards' "genetic introduction to algebraic number theory" **[48]**.

[44] For Kummer's exact words, see **[110]**.

[45] As Karine Chemla **[15]** has recently pointed out, ideal elements had been introduced in the geometrical work of Poncelet. Chasles had provided an explicit roadmap for how to define such ideal elements. Kummer seems to have followed Chasles' approach when introducing ideal elements into arithmetic.

that allows us to detect the presence of $\mathfrak{p}$. See the next chapter for Kummer's choice of $h(\alpha)$.) The key theorems, of course, are (1) that these multiplicities behave as expected and (2) that when ideal prime factors are taken into account one can prove that factorizations exist and are unique.[46] Kummer's theory was developed in a long series of papers that fill most of the first volume **[102]** of his collected papers.

In modern language, we can interpret Kummer's ideal primes in terms of valuations: he gave explicit congruence-based descriptions of all the valuations of the field $\mathbb{Q}(\alpha)$. The fact that such descriptions were possible was crucial. It would not have been acceptable to simply consider all possible valuations, as one might in a modern text.

Two important complications should be noted. First, in the special cases $n = 3$ and $n = 4$ considered by Gauss and Eisenstein, there are only finitely many invertible elements in the ring $\mathbb{Z}[\alpha]$. Once $n \geq 5$ there are infinitely many units, and describing them completely is difficult. Kummer knew how to construct an infinite family of units, now often called the "circular units." Units in the cyclotomic case were the topic of Kronecker's doctoral thesis (1845); Dirichlet published his more general unit theorem in 1846, but the evidence suggests that he had been working on it for a long time.

Second, Kummer's ideal primes only make sense as *factors*.[47] If $\mathfrak{p}$ is an ideal prime, it makes sense to ask whether $f(\alpha)$ is divisible by $\mathfrak{p}$, and therefore it makes sense to as whether there are solutions to $x^n \equiv f(\alpha) \pmod{\mathfrak{p}}$, which allows us to define a degree n symbol $\left(\frac{f(\alpha)}{\mathfrak{p}}\right)_n$. But it does *not* make sense to consider the congruence $x^n \equiv \mathfrak{p} \pmod{f(\alpha)}$, because the difference $x^n - \mathfrak{p}$ is not defined. This makes even the statement of degree n reciprocity difficult, since it presumably must involve flipping degree n symbols.

Kummer's partial solution to the latter difficulty is worthy of note. He first defined equivalence classes of ideal numbers and showed, like Gauss, that the class group is finite. This means that for every ideal prime $\mathfrak{p}$ there is an exponent h such that $\mathfrak{p}^n$ is an actual[48] number: $\mathfrak{p}^h = g(\alpha)$. Then one can try to define

$$\left(\frac{\mathfrak{p}}{f(\alpha)}\right)_n = \left(\left(\frac{g(\alpha)}{f(\alpha)}\right)_n\right)^{1/h}.$$

Since the degree n symbol is an n-th root of unity, this makes sense as long as $\gcd(h, n) = 1$. Kummer defined n to be a *regular prime* if the order of the class group was not divisible by n, in which case this definition always makes sense. The desired application to a general n-th power reciprocity law was therefore only partially successful: Kummer was able to obtain an n-th power reciprocity law when n is a regular prime.

Kummer's cyclotomic theory seems at first to have been regarded mostly as a tool, something used to formulate and prove a more general reciprocity law.[49] From that point of view, the results were somewhat disappointing, with a theorem

---

[46]These results are not at all easy to prove. Indeed, Kummer's proofs were incomplete, as Haubrich **[73]** and Lemmermeyer **[107]** point out. Specifically, what is missing is a proof that if $\mathfrak{p}^m$ divides $f(\alpha)$, then $\mathfrak{p}^{m-1}$ also divides $f(\alpha)$.

[47]Alas, this is sometimes obscured by Kummer's decision to use symbols like $\psi(\alpha)$ for his ideal prime divisors even when they cannot be expressed as polynomials in α.

[48]Kummer used "real number" for elements of $\mathbb{Z}[\alpha]$, which is confusing for modern readers.

[49]As Epple **[50**, I.2, § 7] suggests, in mathematics we can talk about "epistemic objects" and "epistemic techniques" used to understand them, but often the techniques themselves become

that applied only in a special case. It is possible that this is what led Kummer to change his research focus in the 1850s.

One indication that the theory of ideal prime divisors in cyclotomic fields had independent interest was Kummer's proof of Fermat's Last Theorem when the exponent n is a regular prime. Since Kummer did not develop his theory with that application in mind (see, for example, [**42**]), his success in using it to solve an apparently unrelated problem was significant. Here too, however, the restriction to regular primes was disappointing.

Kummer's work clearly served as inspiration and guide for Kronecker and Dedekind. Both referred to Kummer with respect and admiration, but both ended up concluding that a straight generalization of Kummer's approach could not work. The reasons they gave will be discussed in the next chapter.

1.9. A Changing of the Guard

Things changed in the 1850s and 1860s. Jacobi died in 1851, Eisenstein in 1852, Gauss in 1855, and Dirichlet in 1859. Of those mentioned above, only Kummer, Hermite, and Kronecker survived, living until 1893, 1901, and 1891, respectively. Richard Dedekind was younger and lived until 1916.

After around 1860, the coherence and sense of common purpose among number theorists seems to fall apart. One sees

> ... a low point, or at least a decline of the ebullient research field opended at the middle of the century, as its most important contributors left it. This at any rate is the impression conveyed by the contemporary mathematicians themselves. [**65**, p. 70]

Hermite would say in 1882 that "the direction of the mathematical trend ... does not go now toward arithmetic."[50]

This does not mean, of course, that there was no mathematical activity related to number theory. Goldstein and Schappacher show in [**65**], however, that what before was a coherent research field broke up into several "clusters" of activity:

> ... inside each cluster, references to each other's articles and results are frequent, the papers are often close in terms of methods, points of view, or objectives; at least an awareness of the works of others is indicated, while the *mathematical* exchanges between two different clusters of articles are limited, in some cases even tainted by technical misunderstanding or disapproval. [**65**, p. 72]

We refer to [**65**] for a more complete description of these clusters and for a discussion of the need for a more complete analysis of the period.

By the 1850s, Kummer had completed his work on cyclotomic integers. After that, despite occasional returns to number theory, he concentrated his attention on algebraic geometry.

---

objects for further investigation. Kummer sems to have seen himself as creating a new technique for studying reciprocity laws, while Dedekind was clearly interested in the technique itself.

[50]Hermite writing to Smith, quoted by Goldstein and Schappacher in [**65**, p. 71]. The context is the lack of interest in "a prize posted by the Paris Academy on the representation of integers as sums of five squares." Hermite writes "Jusqu'ici je n'ai pas eu connaissance qu'aucune pièce ait été envoyée, ce qui s'explique par la direction du courant mathématique qui ne se porte plus maintenant vers l'arithmétique."

Between 1848 and 1855, Kronecker was away from Berlin on family business, which must also have slowed him down. He returned to Berlin in 1855 and resumed his mathematical work, first as an independent scholar, then as a member of the Academy of Sciences, and finally, in 1883, as a professor at the University of Berlin.[51]

Richard Dedekind began his mathematical life in this period, completing his thesis under Gauss in 1852 and becoming Dirichlet's devoted follower after 1855. The deaths of Gauss and Dirichlet seem to have had a large impact on Dedekind which was exacerbated by the death of his close friend Bernhard Riemann, also quite young, in 1866. Dedekind spent much of the next decade working on preserving the work of his friends and teachers. He was involved in the production of the first two volumes of Gauss's *Werke*. His write-up of Dirichlet's lectures on number theory also appeared in 1863, with a second edition in 1871 and two further editions after that. Riemann's collected works, published in 1876, were edited by Dedekind and Heinrich Weber.

In Berlin, Kronecker was near the center of German mathematics. He worked on a wide range of subjects and had many doctoral students. The "triumvirate" of Kummer, Kronecker, and Karl Weierstrass dominated mathematics in Berlin. Dedekind, by contrast, settled in Braunschweig and seems to have been content to work mostly on his own. He seems to have had no doctoral students.

1.10. Hints of a Theory of Algebraic Integers

In hindsight it is easy to assume that once Kummer had developed a good theory of factorization in the cyclotomic case it would have been immediately apparent that the general case should be studied. There is very little evidence, however, of any desire for a general theory.

A sense of the typical attitude can be seen in the second part (1860) of Smith's *Report on the Theory of Numbers*. After giving an account of Kummer's theory of cyclotomic integers, Smith argues explicitly that there is no need for a general theory of algebraic numbers:

> It will be remembered that the complex numbers to which our attention has been directed are not of that general kind to which we have referred in Art. 41, but are exclusively those which are composed of roots of unity. The theory of complex numbers, in the widest sense of that term, does indeed present to us an important generalisation of the theory of the residues of powers; for the theorem of Fermat[52] (see Art. 53) subsists alike for every species of complex numbers. But the complex numbers of Gauss, of Jacobi, and of M. Kummer force themselves upon our consideration, not because their properties are generalisations of the properties of ordinary integers, but because certain of the properties of integral numbers can only be explained by a reference to them. [**129**, Art. 64]

[51]Kummer retired in 1883 and Kronecker was appointed professor as his replacement. Thus, in a formal sense, Kronecker was the successor of Kummer.

[52]Smith means "Little Fermat," which for cyclotomic integers takes the form $f(\alpha)^{N-1} \equiv 1 \pmod{\mathfrak{p}}$ when $N = N(\mathfrak{p})$ and $\mathfrak{p}$ does not divide $f(\alpha)$.

For Smith, higher reciprocity laws were the motivation and the only justification for studying "complex numbers," and only cyclotomic complex numbers were needed for that purpose.

Dirichlet was something of an exception; more than others at the time he was interested in general "complex numbers." His unit theorem is one of the more striking examples (see [**111**, Section 13.2]). Let α be a root of a monic irreducible polynomial with integer coefficients; we can then study polynomials in α with integer coefficients, what we now call the ring $\mathbb{Z}[\alpha]$. There are usually many invertible elements, known as "units," in this ring. (Dirichlet actually phrased this in terms of the norm form, but he did use the term "complex units.") There are usually infinitely many units, but they are generated by a finite number among them; Dirichlet proved this and determined the exact number of generators. For polynomials of degree two with real roots, this theorem boils down to the Pell Equation, but for all higher degrees even the fact that there were infinitely many solutions was new. When Kummer started working on the cyclotomic theory, he found infinitely many units (what we now know as the circular units), but he seems to have deferred to Dirichlet about the details. In any case, Dirichlet's unit theorem seems to have been one of the few important results about general algebraic integers known before the work of Dedekind and Kronecker.[53]

Few other mathematicians seemed interested in generalizing Kummer's cyclotomic theory. Nevertheless, there were some hints that more general algebraic integers were worth investigating. We highlight some of those hints here. As we will see in Chapter 3, however, as late as 1871 Dedekind lamented the general lack of interest in the subject (see Section 3.4).

Perhaps the most significant hint that more general algebraic numbers could be useful came in Kummer's work. To formulate and prove his degree n reciprocity law (for prime n), Kummer worked with algebraic integers formed from an n-th root of unity. To push the proof through, however, he had to consider n-th roots of his algebraic integers, which of course do not usually belong to the cyclotomic field. In modern language, Kummer was working in $K = \mathbb{Q}(\zeta_n)$ for a prime number n, but needed to use the larger field $K(\sqrt[n]{u})$. (Such fields are now known as Kummer extensions.) So already in Kummer's theory there is a pointer to the need for more general fields.[54] When $n = 2$, we have $K = \mathbb{Q}$, so that Kummer's results already included the arithmetic of quadratic fields (except for the primes dividing twice the discriminant).[55] Kummer hinted at the possibility of interpreting the theory of binary quadratic forms in terms of quadratic fields but the idea was not developed further until Dedekind's work.

Another big, but more obscure, hint came from the theory of elliptic functions with complex multiplication. These are connected to imaginary quadratic fields, which therefore appear naturally in the theory of positive-definite binary quadratic forms. (These connections fill Part VI of Smith's report [**129**].) The values taken

[53]But note that Dirichlet always worked with $\mathbb{Z}[\alpha]$ rather than the ring of all algebraic integers.

[54]Kummer's treatment of these fields avoids the primes that divide nu, i.e., those that divide the discriminant of the polynomial $x^n - u$, showing that he was to some extent aware of the difficulties outlined in the next chapter. See [**73**, IV.1.2] for more details.

[55]As Haubrich observes, "Kummer himself never said a word about this special case." ("Kummer selbst hat mit keinem Wort auf den Spezialfall hingewiesen." [**73**, p. 67]

by certain modular functions at singular moduli were algebraic integers of a particularly interesting type. Understanding exactly what was going on here took a long time.[56]

In the 1850s, Kronecker had announced his theorem about abelian equations[57] with rational coefficients. Kronecker's claim[58] was that the roots of any such equation could be expressed in terms of roots of unity.

Kronecker seems to have realized that the polynomials associated with elliptic and modular functions with complex multiplication might play the analogous role for abelian equations whose coefficients were in an imaginary quadratic field. He described this as his "Jugendtraum," the dream of his youth, which he never quite managed to prove (but which led to a great deal of important mathematics). The roots of these polynomials were algebraic integers which it was natural to study more deeply.[59] Elliptic functions were a main focus of Kronecker's mathematical work, so it is not surprising that he tells us that this was his motivation for returning to the study of algebraic numbers:

> Later in the years 1856 and 1857 I was prompted by the study of the complex multiplication of elliptic functions to return to my earlier investigations on complex numbers formed from roots of arbitrary polynomial equations $F(x) = 0 \ldots$[60] [**97**, § 19, p. 325]

Apart from Kronecker, there was a small group of mathematicians in Dedekind's circle who seemed interested in general "complex numbers." Eduard Selling and Paul Bachmann published some early attempts,[61] but the first to publish a fully successful general theory was Dedekind. We give in Chapter 3 a translation of his announcement. E. I. Zolotarev's theory was mostly ignored; we discuss some of the reasons in Chapter 4. Kronecker's version was incorporated into his more general *Grundzüge*, which we discuss briefly in Chapter 5. Proofs of several of Kronecker's results were finally published by Hensel; we discuss his work in Chapter 6. All of this work was conditioned by the problem of common inessential discriminant divisors, which we explain in the next chapter.

---

[56]The theory is still hard to understand. See [**138**] for a brief history; see the latter part of [**17**] for a modern account.

[57]In modern terms, polynomials whose splitting field had abelian Galois group. Kronecker's original theorem focused on the case where the Galois group is cyclic.

[58]Kronecker never did fully prove this theorem, which is now known as the Kronecker-Weber theorem.

[59]Hilbert eventually stated a (wrong) version of Kronecker's Jugendtraum as the twelfth of his problems for the new century. See [**124**] for this story. [**138**] is a more complete account of complex multiplication and the Jugendtraum. [**17**] gives the modern form of the theory.

[60]Als ich dann später in den Jahren 1856 und 1857 durch das Studium der complexen Multiplication der elliptischen Functionen veranlasst wurde, auf meine früheren Untersuchungen über complexe aus Wurzeln beliebiger ganzzahliger Gleichungen $F(x) = 0$ gebildete Zahlen zurückzukommen...

[61]See Haubrich's discussion in [**73**, ch. IV] and the next chapter.

Names and Dates[62]

Joseph-Louis Lagrange, 1736–1813

Adrien-Marie Legendre, 1752–1833

Carl Friedrich Gauss, 1777–1855

Niels Henrik Abel, 1802–1829

Carl Gustav Jacobi, 1804–1851

Peter Gustav Lejeune Dirichlet, 1805–1859

Ernst Eduard Kummer, 1810–1893

Évariste Galois, 1811–1832

Charles Hermite, 1821–1901

Gotthold Eisenstein, 1823–1852

Leopold Kronecker, 1823–1891

Richard Dedekind, 1831–1916

Eduard Selling, 1834–1920

Paul Bachmann, 1837–1920

Egor Ivanovich Zolotarev, 1847–1878

Kurt Hensel, 1861–1941

[62]Main source: *The Complete Dictionary of Scientific Biography* [**51**].

CHAPTER 2

An Outline of the Problem

The main goal of this chapter is to set out, mostly in modern language, the "technical difficulty" that we want to discuss. While we make some historical remarks, we do not aim for a complete account, which would require much more historical data.

2.1. Generalizing Kummer

We know only a few mathematicians who thought about how to generalize Kummer's theory of "complex numbers built from roots of unity." Kummer tells us[1] that Leopold Kronecker had some sort of theory in 1859. Eduard Selling published in 1865 an attempt to extend Kummer's theory to Galois number fields, see **[127]**. His paper is difficult to understand and seems not to have convinced anyone. Paul Bachmann published a short book **[4]** about "complex numbers" of the form $a + b\sqrt{D} + c\sqrt{\Delta} + d\sqrt{D\Delta}$ in 1867. (Many mathematicians may have seen only the shorter account **[5]** published in Crelle's Journal, which had very few details.) Both Selling and Bachmann had attended Dedekind's seminar on Kummer's cyclotomic theory, as Dedekind tells us,[2] so their work may reflect his influence.[3]

Richard Dedekind was the first one to go to print with a satisfactory theory, in 1871, as we will discuss in Chapter 3. In 1874, Zolotarev published a paper **[143]** attempting a simpler theory, prompting the response from Dedekind that we translate in Chapter 4. A little later, Zolotarev did achieve a satisfactory theory, which became accessible to a wider range of mathematicians when it appeared in French (see **[144]**) in 1880. We discuss this in more detail in Chapter 4.

Kronecker published a vast generalization in **[97]**, which appeared in 1882. The various theories remained in competition at least until the publication of Hilbert's *Zahlbericht* **[87, 89]**, which surveyed all the work done in the 19th century and set the terms for the work that followed.

As all of them tell us, the natural place to start was the theory of "higher congruences," but, as we explain in this chapter, there are several obstacles in the way of a generalization. Both Dedekind and Kronecker decided, in the end, that these obstacles meant that the approach via congruences was the wrong one. Each invented a new technique: ideals in Dedekind's case, the adjunction of indeterminates in Kronecker's. Each later argued, as we will see in this book, that it was *impossible* to proceed via congruences, using this claim to answer the natural question: "must it be so complicated?"

---

[1] In **[103**, p. 57]; see the beginning of Chapter 5.

[2] See his prefatory remarks in Section 3.3 (on page 34).

[3] Though his major goal is to discuss Dedekind's path to his theory of algebraic integers, Haubrich discusses these early attempts in chapter IV of **[73]**.

2.2. Higher Congruences and Factoring Primes

As we described in Section 1.8, Kummer's theory concerned "complex numbers" of the form

$$a_0 + a_1\zeta + a_2\zeta^2 + \cdots + a_{n-1}\zeta^{n-1},$$

where $a_i \in \mathbb{Z}$ and ζ is a complex n-th root of unity, so that $\zeta^n = 1$ and $\zeta^k \neq 1$ for $1 \le k < n$. Such numbers are now called *cyclotomic integers* and the ring they form is denoted $\mathbb{Z}[\zeta]$. Kummer's theory was developed mostly under the assumption that n is a prime number; though he later generalized his results, we will stick to that assumption in this section.

Since n is prime, the minimal polynomial with integer coefficients satisfied by ζ is

$$\Phi_n(x) = \frac{x^n - 1}{x - 1} = 1 + x + x^2 + \cdots + x^{n-1}.$$

This is known as the n-th *cyclotomic polynomial.*

Kummer discovered that $\mathbb{Z}[\zeta]$ does not usually have the unique factorization property,[4] but had the fundamental insight that he could "save" unique factorization by allowing more factors, which he called "ideal prime factors." For this to work, it was essential[5] to have a method to establish exactly when a cyclotomic integer was divisible by an ideal prime factor.

All of the founders of algebraic number theory saw the most potential for generalization in the following theorem, which seems to be only implicit in Kummer. It holds for all integers $n > 2$, but we stay with the main case and state it only for n prime.

THEOREM 2.2.1. *Let $n > 2$ be a prime number, let ζ be a primitive n-th root of unity, and let $\Phi_n(x)$ be the n-th cyclotomic polynomial. Fix a prime number $p \in \mathbb{Z}$ and let*

$$\Phi_n(x) \equiv F_1(x)^{e_1} F_2(x)^{e_2} \ldots F_r(x)^{e_r} \pmod{p}$$

be the factorization of $\Phi(x)$ modulo p, where the $F_i(x)$ are distinct irreducible polynomials in $\mathbb{F}_p[x]$. Then the factorization (up to units) of (p) in $\mathbb{Z}[\zeta]$ is

$$(p) = \mathfrak{p}_1^{e_1}\mathfrak{p}_2^{e_2} \ldots \mathfrak{p}_r^{e_r},$$

with distinct ideal prime factors $\mathfrak{p}_i = \gcd(p, F_i(\zeta))$.

When $p = n$ it is easy to see that the factorization is $\Phi_n(x) \equiv (x-1)^{n-1}$, so $(n) = (1-\zeta)^{n-1}$ up to unit factors. When $p \neq n$ we always have $e_i = 1$ and the factorization depends only on the least integer f such that $p^f \equiv 1 \pmod{n}$.

It was particularly important for Kummer's theory that the polynomials $F_i(x)$ could be found explicitly and used to create a congruential test for divisibility by $\mathfrak{p}_i$. Assume $p \neq n$, so that all the exponents $e_i = 1$. Then with the above notation we have

$$(p) = \mathfrak{p}_1\mathfrak{p}_2 \ldots \mathfrak{p}_r.$$

Suppose we want to test a complex number $g(\zeta) \in \mathbb{Z}[\zeta]$ for divisibility by $\mathfrak{p}_1$. We consider the polynomial

$$H_1(x) = F_2(x) \ldots F_r(x)$$

---

[4]Unique factorization fails for all prime $n \geq 23$.

[5]See **[15]** for a discussion of how nineteenth century mathematicians would justify the introduction of "ideal elements" in mathematics.

obtained by multiplying all but the first factor, and look at $g(\zeta)H_1(\zeta)$. If each coefficient of this complex number is divisible by p, then we say that $g(\zeta)$ is divisible by $\mathfrak{p}_1$. In other words, the number $H_1(\zeta)$ is divisible exactly once by all of the other ideal prime divisors of p, and so multiplying by it provides a way of testing for divisibility by $\mathfrak{p}_1$. To test for divisibility by a power of $\mathfrak{p}_1$ we use a power of $H_1(\zeta)$ instead. The hard work goes into proving that this defines a valuation on $\mathbb{Z}[\zeta]$.

2.3. Can We Generalize?

It was the simple description of factorization in terms of congruences that seemed to suggest the possibility of a very simple theory in the general case: for a general number field $\mathbb{Q}(\alpha)$, let $\Phi(x)$ be the minimal polynomial for α; let's say its degree is k. Then we factor $\Phi(x)$ modulo p and use this to define "ideal primes" à la Kummer.

The choice of α is the first problem. Many different equations produce closely related "complex numbers." Consider, for example, the complex numbers defined by the equations $\rho_1^2+3=0$, $\rho_2^2+\rho_2+1=0$, and $4\rho_3^2+3=0$. Each can be expressed in terms of the others: $\rho_2=(-1+\rho_1)/2$, $\rho_3=\frac{1}{2}\rho_1$. Kronecker would say that these algebraic numbers are all of the same "genus" (Gattung). Dedekind would say that they generate the same field.

If we try to study the arithmetic of the corresponding "complex numbers", however, the three choices give very different results. Working with numbers of the form $a+b\rho_3$ doesn't work at all: the product of two such numbers is not a number of this form (consider ρ_3^2). The other two at least make sense: both $\mathbb{Z}[\rho_1]$ and $\mathbb{Z}[\rho_2]$ are rings. But the fact that $(\rho_1-1)^3=8=2^3$ shows[6] that there can be no theory of ideal prime divisors for $\mathbb{Z}[\rho_1]$. On the other hand, ρ_2 is a cube root of unity and so Kummer's theory applies directly.

Kummer seems to have been aware of this difficulty. In **[103]** he considered complex numbers involving a λ-th root of unity α, with λ a prime number, together with a number u such that $u^\lambda=D(\alpha)$. He says that he will not try to find ideal prime factors in $\mathbb{Z}[\alpha,u]$ of the primes in $\mathbb{Z}[\alpha]$ that divide $\lambda\, D(\alpha)$, not only because he does not need them but also because they "in general cannot be decomposed into true ideal prime factors."[7]

What distinguishes these "bad" primes from the rest is that they are divisors of the discriminant of the defining polynomial. We see a similar thing in Bachmann's work **[4, 5]** on biquadratic numbers of the form $a+b\sqrt{D}+c\sqrt{\Delta}+d\sqrt{D\Delta}$: primes dividing $2D\Delta$ are "abstrahirt."

Is there a "best" generator? The example of ρ_3 above shows that we want a root of a *monic* irreducible polynomial. The fact that the discriminant of x^2+3 is -12, and that of x^2+x+1 is -3 suggests that one idea is to choose a polynomial with minimal discriminant, but that does not turn out to work either.

---

[6] The identity shows that ρ_1-1 and 2 must have exactly the same prime factorization up to units, and hence must differ by a unit. But the only units in $\mathbb{Z}[\rho_1]$ are 1 and -1.

[7] "... im allgemeinen in wahre ideale Primfaktoren sich gar nich zerlegen lassen, welche im vollen Sinne diesen Namen verdienen."

Kronecker and Dedekind both ended up with the same crucial insight.[8] In modern terms, which are closer to how Dedekind would say things, it is this.[9] First define α to be an *algebraic integer* if it is a root of a monic irreducible polynomial with integer coefficients. Next, consider all the elements in the field $\mathbb{Q}(\alpha)$ that are algebraic integers. Dedekind and Kronecker proved that the sum and product of two algebraic integers are still algebraic integers, so that the algebraic integers form a ring $\mathcal{O} \subset \mathbb{Q}(\alpha)$. Both of them also showed that it is a free $\mathbb{Z}$-module, i.e., that one can find k numbers

$$\omega_1, \omega_2, \ldots, \omega_k \in \mathcal{O}$$

such that:

(a) Every element in the field $\mathbb{Q}(\alpha)$ can be written uniquely as a $\mathbb{Q}$-linear combination of the ω_i. (So k is the dimension of $K = \mathbb{Q}(\alpha)$ over $\mathbb{Q}$.)
(b) An element $\theta \in \mathbb{Q}(\alpha)$ is an algebraic integer if and only if it is an *integral* linear combination of the ω_i. (So the ω_i are a basis of $\mathcal{O}$ as a free $\mathbb{Z}$-module of rank k.)

Both of them used the "fundamental series" ω_i to define a number called the *discriminant* of the field $K = \mathbb{Q}(\alpha)$. For each i we consider the conjugates of ω_i (the other roots of its minimal polynomial). Write them as $\omega_i^{(1)} = \omega_i, \omega_i^{(2)}, \ldots, \omega_i^{(j)}$. Then the discriminant d_K is defined as the square of the determinant of the matrix $[\omega_i^{(j)}]$.

Once we know we must work with the full ring $\mathcal{O}$ we can restate the first difficulty: in the case of $\mathbb{Q}(\zeta)$ the ring of algebraic integers is exactly $\mathbb{Z}[\zeta]$, but in general one cannot expect that there will be a single generator. If K is a number field and $\mathcal{O} \subset K$ is its ring of algebraic integers there may not exist *any* $\alpha \in \mathcal{O}$ such that $K = \mathbb{Q}(\alpha)$ and $\mathcal{O} = \mathbb{Z}[\alpha]$.[10] In such a situation, there is no obvious $\Phi(x)$ to work with.

One way to detect that $\mathbb{Z}[\alpha]$ is not equal to the full $\mathcal{O}$ is to compare the discriminant of the irreducible polynomial α satisfies with the discriminant of the field. If they are equal, then $\mathbb{Z}[\alpha] = \mathcal{O}$. If not, the polynomial discriminant will be a multiple of the field discriminant.

The subring $\mathbb{Z}[\alpha]$ of the full ring $\mathcal{O}$ of algebra integers is an example of what Dedekind called an *order*. The maximal order in the field K is exactly $\mathcal{O}$. Kronecker would call $\mathbb{Z}[\alpha]$ a *species* (Art) and $\mathcal{O}$ the principal species (Haupt-Art). If one tries to do arithmetic in a non-maximal order one is forced to exclude certain primes.

As Haubrich points out in **[73]**, Kummer was extremely lucky. Haubrich lists (see p. 58) five special features of the cyclotomic integers $\mathbb{Z}[\zeta]$, all of which fail in the general case. We give them in modern language:

(a) $\mathbb{Z}[\zeta]$ is the maximal order in $\mathbb{Q}(\zeta)$, i.e., it is already the ring of all algebraic integers in $\mathbb{Q}(\zeta)$.

[8] H. M. Edwards argues that once a correct definition of "integer" is in hand the generalization of Kummer's theory "is not difficult," [**43**, p. 337]. He does not discuss CIDDs, which both Dedekind and Kronecker say were the real difficulty.

[9] It is unclear what led them to this insight. Edwards says "Unfortunately, neither of them gives a reason for this definition or explains how it was arrived at." [**43**, p. 332]. According to Haubrich [**73**, pp. 70–71], Selling had noticed that there were numbers that should count as "integers" but were not integer linear combinations of the roots of his defining polynomial. He could not, however, formulate a simple criterion for which numbers they were.

[10] Indeed, one expects that it is almost never the case except for quadratic and cyclotomic fields. See, for example, [**1**].

(b) $\mathbb{Q}(\zeta)$ is normal and its Galois group over $\mathbb{Q}$ is abelian. In the case of prime n, it is even cyclic.
(c) The conjugates of ζ are all powers of ζ. Indeed, the powers of ζ form a normal integral basis.
(d) When n is prime there is only one prime divisor of n in $\mathbb{Z}[\zeta]$ and the factorization is simple to give, as we saw above.
(e) Factoring the cyclotomic polynomial modulo p yields the factorization of p in $\mathbb{Z}[\zeta]$.

Some of these are direct consequences of others, so there is some redundancy here, but the list shows that the cyclotomic case is very special. The case of quadratic fields is similarly special. Having started with those cases, Kronecker and Dedekind found themselves in a completely unexpected situation once they started looking at more general fields.

2.4. Dedekind's Factorization Theorem

How can we factor primes $p \in \mathbb{Z}$ when the ring of algebraic integers is not monogenic, that is $\mathbb{Z}[\alpha]$ is a proper subring of $\mathcal{O}$ for *every* algebraic integer α in our field? It turns out that we can get away with a weaker assumption.

The first step is to compare the sizes of $\mathbb{Z}[\alpha]$ and the bigger ring $\mathcal{O}$. Both have a basis over $\mathbb{Z}$. For $\mathbb{Z}[\alpha]$ it is just the powers of α; for $\mathcal{O}$ it is the integral basis ω_i. Since $\mathbb{Z}[\alpha]$ is contained in $\mathcal{O}$, we can express each power of α as a linear combination of the ω_i. This gives us a $k \times k$ matrix; if its determinant is ± 1 then $\mathbb{Z}[\alpha] = \mathcal{O}$, but in general the determinant is larger. The absolute value of the determinant is what Dedekind called the *index* of α. In modern language it is the index of $\mathbb{Z}[\alpha]$ as a subring of $\mathcal{O}$.

We can now state the result that has become known as "Dedekind's Theorem," which generalizes the result given above in the cyclotomic case. Dedekind announced this result in 1871 (see Chapter 3) and published a proof in 1878 (see Chapter 4). It seems clear that Kronecker was also aware of some form of this theorem.

THEOREM 2.4.1. *With the notations above, given a prime number $p \in \mathbb{Z}$, suppose we can find an α such that $\mathbb{Z}[\alpha] \subset \mathcal{O}$ has index not divisible by p. Then factoring the minimal polynomial for α modulo p gives the correct factorization of (p) in $\mathcal{O}$.*

This allowed one to hope, then, that an explicit factorization theory could be based on a "local" approach: for each prime p, find a generator α such that p does not divide the index $(\mathcal{O} : \mathbb{Z}[\alpha])$. Then apply the theorem to find the factorization. Kronecker seems to have had something like this in mind when he said "I tried at first to eliminate the difficulty presented by the inessential prime factors of the discriminant ... by using other equations defining the same genus."[11] Dedekind says in [**20**, p. 1490] at first he thought it "very likely"[12] that for each prime it would be possible to find such an α (see p. 39). But, he says, "all my attempts

[11]Die Schwierigkeit, welche die ausserwesentlichen Primfactoren der Discriminante ... suchte ich Anfangs dadurch zu beseitigen, dass ich andere Gleichungen derselben Gattung zu Grunde legte." [**97**, § 19]

[12]sehr wahrscheinlich

to prove the existence of such a number were unsuccessful."[13] Indeed, Dedekind's hope was not to be fulfilled: there exist number fields in which *all* of the indices have a common prime divisor. Dedekind pointed this out in his book notice [**20**], translated in Chapter 3, probably to explain why he had needed a new approach. Kronecker says in his *Grundzüge* [**97**, §25, p. 384] of 1882 that he had found an example of the same phenomenon in 1858. We give that text in Chapter 5.

Dedekind and Kronecker, and later Hensel, pointed to this essential difficulty to justify introducing a new approach: ideals in Dedekind's case, polynomials in many variables in Kronecker's. Zolotarev tried to extend Kummer's theory directly in this style [**143**], but he knew that this method would fail for finitely many primes. (Eventually, in a second paper [**144**], Zolotarev found still another way to work around the difficulty.) Dedekind's paper [**23**], translated in Chapter 4, was prompted by an announcement of Zolotarev's work.

Kronecker stated the problem in terms of discriminants. For each choice of α, let $d(\alpha) = \operatorname{disc}(\Phi(x))$ be the discriminant of its minimal polynomial. Let d_K be the field discriminant. Then $d(\alpha) = m^2 d_K$, where m is exactly the index of $\mathbb{Z}[\alpha]$ in $\mathcal{O}$. Therefore the many polynomial discriminants $d(\alpha)$ have a common factor d_K, which Kronecker called the *essential* part, attached to the field (or "genus domain") rather than to a specific element. The other factors of $d(\alpha)$ (i.e., the factors of m) are therefore "inessential."[14] So in the "bad" examples what is happening is that some prime p is an inessential divisor of every element discriminant. Such primes were the "common inessential discriminant divisors."

The name is perhaps ill-chosen, because it is perfectly possible for a prime p to divide the discriminant d_K and *also* divide the index of some α.[15] Such a prime divisor is then both "essential" (it divides d_K) and "inessential"! Dedekind's term "index divisor" seems more appropriate. In [**65**] the authors use the term "parasitic factors of the polynomial discriminant." Nevertheless we find Kronecker's term charming and so will use it in this book. We will often use the abbreviation CIDD.

In the later literature, the "index $i(K)$ of the field K" was defined to be the greatest common divisor of the indices of all the generators of K; then Kronecker's common inessential discriminant divisors are just the divisors of $i(K)$. See [**113**, 2.2.1 item 3] for information on more recent work.

2.5. Give Me an Example

In the next chapter, we will see Dedekind's explanation of how one might look for an example and his careful construction of a particularly simple one. Indeed, Dedekind shows that his theorem above *forces* the existence of many examples.

Kronecker said in the *Grundzüge* that he found an example "in the thirteenth roots of unity" already in 1858 (see the translation in Chapter 5). He never gave the details, but they are probably as Hensel gave them in his Ph.D. thesis[16] [**74**]. Here is the example, in modern language.

---

[13]alle meine Versuche, die Existenz einer solchen Zahl nachzuweisen, fruchtlos blieben [**23**, p. 23]; see the opening paragraph in Section 4.3, §4, page 63.

[14]Kronecker's term was term was "ausserwesentlichen Discriminantentheiler", which is literally something like "extra-essential discriminant divisors."

[15]See, for example, global field 4.0.2156.2 in [**109**] and the second example in Section 6.5.1.

[16]Our decription of Hensel's thesis follows Petri's account in [**117**, 2.2], which takes into account information from Hensel's mathematical diary from that period.

Let ζ be a primitive 13-th root of unity. Gauss showed that there is a unique subfield K of degree 4 over $\mathbb{Q}$.[17] Since the discriminant of $\mathbb{Q}(\zeta)$ is a power of 13, so is the discriminant of K (in fact, $d_K = 13^3$).

We will show that the prime 3 divides the discriminant of the minimal polynomial of any element $\alpha \in \mathcal{O}$. Notice first that the order of 3 modulo 13 is three, i.e., $3^3 \equiv 1 \pmod{13}$ but no smaller power of 3 is congruent to 1. It follows from Kummer's work that the prime number 3 is divisible by four ideal primes in K, each of which has norm 3; let $\mathfrak{p}$ be one of these.

Since $N(\mathfrak{p}) = 3$, the field $\mathcal{O}/\mathfrak{p}$ has three elements (or, as Kronecker and Dedekind would say, there are three congruence classes modulo $\mathfrak{p}$). Consider an algebraic integer $\alpha \in \mathcal{O} \subset K$ such that $K = \mathbb{Q}(\alpha)$. Since K is a normal field of degree 4, the discriminant of the minimal polynomial of an integer in K is the square of the product of the differences of four integers in K, the four roots of the minimal polynomial.[18] Since there are only three congruence classes of integers modulo $\mathfrak{p}$, at least two of those four integers must be congruent modulo $\mathfrak{p}$, so that at least one of the differences must be divisible by $\mathfrak{p}$. Hence $d(\alpha)$ will be divisible by $\mathfrak{p}$. Since $\mathfrak{p}$ lies above 3 and $d(\alpha)$ is an integer, it must in fact be divisible by 3. The field discriminant d_K, however, is a power of 13; thus, 3 is an inessential divisor. The whole analysis is true for any integral generator α, so 3 is a common inessential discriminant divisor.

Clearly this idea can be generalized: whenever the field K is normal and the k roots of the minimal polynomial for α need to be fit into fewer than k congruence classes mod $\mathfrak{p}$, the difference between two of them will be divisible by $\mathfrak{p}$ and hence $d(\alpha)$ will be divisible by p. This generalization seems to have been the initial focus of Hensel's Ph.D. work (see [**117**, 2.2]).

2.6. What Next?

The first role of common inessential discriminant divisors was as a sign that a certain route did not lead to the desired goal: polynomial congruences cannot be used to create a simple generalization of Kummer's cyclotomic theory. One could then either attempt to create a theory on a different basis or use congruences when they worked and supplement them when necessary. According to a story Dedekind seems to have told,[19] Kummer thought the second method was better; it was essentially what Zolotarev did. Dedekind and Kronecker wanted a uniform

[17]This is global number field 4.0.2197.1 in the LMFDB database [**109**].

[18]If K were not normal, the conjugates might not be in K, and the argument falls through.

[19]See [**25**, vol. 3, p. 481]. The story comes from the notes taken by Felix Bernstein after a conversation with Dedekind in March 1911. Bernstein quotes Dedekind telling of a visit to Kummer:

> Kummer, by the way, did not want to know anything at all about my investigations. When I came to Berlin to visit my friend H. Weber, I also went to see Kummer, and he did not receive me kindly at all. He said: "You are probably coming to see if I am not departing soon." I said: "I have come to see the man whom I admire most and to whom I owe the greatest inspiration of my life." Then he became a little more friendly, and in the end we got along quite well. He reproached my theory mainly for what is actually, strictly speaking, its advantage, namely, that I treat the ideals that go into the discriminant completely like the others: "they are something totally different, one must not confuse them."

approach that would treat all primes the same way: Dedekind invented ideals and Kronecker invented the method of adjoining indeterminates to create a theory of divisors.[20] We will describe each approach briefly in the introductions to Chapters 3 and 5.

The second role of this phenomenon was apologetic. To those who disliked the abstractness of ideals or complicated computations with polynomials in many variables, one could answer that the complexity was required by the situation. Kronecker, for example, says that his method accords with "the nature of things" (see the third paragraph of Section 5.4). If the story of Kummer's comments is correct, this is a point where the two friends disagreed: Kummer thought that treating the divisors of the discriminant separately was more natural, while Kronecker (and Dedekind) felt that to capture the essential situation required a theory that treated all primes uniformly.

There were mathematical questions to resolve as well. The most obvious one was to characterize when a prime is a common inessential discriminant divisor. Is it a matter of counting, as in Kronecker's example, or are there deeper reasons? As we will see, the answer to this question was published twice: by Dedekind in 1878 (Chapter 4) and by Hensel in 1894 (Chapter 6).[21] The later paper references the earlier, but there it seems unclear to what extent Hensel knew and understood Dedekind. (See our comments in Chapter 6.)

Perhaps it is a sign of how little Kronecker followed Dedekind's work that he suggested the problem of common inessential discriminant divisors to Hensel for his Ph.D. in 1882, four years after Dedekind's paper. In his thesis, Hensel stated but did not prove the correct criterion.[22] (The same criterion Dedekind had found in 1878.) Petri believes[23] that Hensel had a proof by the time of his Habilitation in 1886, so well before he finally published it in 1894.

One goal of using polynomial congruences and Dedekind's theorem was to factor a rational prime p in the ring of integers $\mathcal{O}$. So there is also an algorithmic[24] question. Suppose I have a prime p. Can I tell whether Dedekind's theorem applies or not? If it does not apply, how do I go about factoring p? The characterization of CIDDs that both Dedekind and Hensel eventually found depends on knowing how p factors, so it does not fully resolve the issue: if we know the factorization of p, then we can decide if we can apply Dedekind's theorem, but we cannot find the factorization except by using Dedekind's theorem. As we will see, Hensel eventually

[20] It's hard to resist: from Kummer's "ideal divisors" Dedekind kept the "ideals" and Kronecker the "divisors."

[21] Hasse, in [**72**, p. 456], attributes this criterion to Hensel and says it was the first success of Hensel's new methods, presumably meaning p-adic methods. In fact the criterion was first found by Dedekind and, as we shall see, neither author used p-adic methods.

[22] Hensel's thesis booklet [**74**] is a curious document, consisting of an introduction outlining an extensive list of theorems followed by proofs of only the first few of those theorems; see the description by Birgit Petri in [**117**, 2.2]. One of the theorems Hensel actually proves is a generalization of the numerical condition in Kronecker's example, It gives a sufficient criterion for the existence of common inessential discriminant divisors. Kronecker emphasized this result, which produced an infinite number of examples, in his report [**10**, ch. 11, Dok. 14] on Hensel's thesis.

[23] Petri, [**117**, 2.4.2].

[24] Despite Dedekind's reputation for ignoring such issues, it is clear that he does want to answer the question of "how do I do it?" So, of course, do Kronecker and Hensel. None of our authors, however, seems very interested in the *efficiency* of their algorithms. For example, applying Dedekind's theorem requires factoring a polynomial modulo p, a problem for which the best available deterministic algorithm runs in time $O(\sqrt{p})$.

found a way to decide whether a primce is a common inessential discriminant divisor without finding its complete factorization; we will discuss his result in Chapter 6.

CHAPTER 3

Dedekind's *Anzeige*, 1871

The second edition of Dirichlet's *Vorlesungen über Zahlentheorie* **[35]** appeared in 1871. The first edition, which was in fact written by Dedekind based on his notes from Dirichlet's lectures, had contained nine supplements, mostly taken from Dirichlet's papers and supplementary lectures.[1] The second edition included a new (tenth) supplement, entitled "On the Composition of Binary Quadratic Forms" **[21]**. Most readers would have expected to find here a (simplified, perhaps) account of the standard theory. If they read it, however, they would have been surprised to find, in the middle of the supplement, a whole new theory of factorization in general fields of algebraic numbers. This new theory was presented in terms of several new ideas: fields, rings of algebraic integers, modules, and ideals.

As he had done for the first edition in 1863, Dedekind wrote an article **[20]** for the September 20, 1871 issue of the *Göttingische gelehrte Anzeigen*. This was a weekly journal containing mostly book reviews, but Dedekind is of course writing about his own book(s). As one would expect, the article (much longer than what he had written for the first edition) focuses almost entirely on the new content, i.e., Supplement X. It gives a cursory summary of the parts of the supplement containing well known material, then discusses the new theory of ideals in more detail.

In the middle of this summary, Dedekind's book notice suddenly swerves into a discussion that goes well beyond what is found in Supplement X. In this digression, Dedekind announced results about the relationship of ideal theory and "higher congruences." He also announced that the prime divisors of the field discriminant were exactly the ramified primes. These theorems were not in Supplement X. Dedekind would only publish the proofs of the theorems on higher congruences in 1878 (**[23]**, translated in the next chapter). The theorem on ramified primes was published even later, in 1882 **[24]**. After this long digression, Dedekind went back to a section-by-section discussion of the Supplement, without giving many details.

It seems that the discussion of "common index divisors" (the name came later) was included in the book notice exactly to explain why the approach via congruences was, as outlined in the previous chapter, bound to fail. This suggests that Dedekind was well aware that his contemporaries would wonder whether something as innovative as the theory of ideals was justified. His digression into "higher congruences" makes the point that a new method is needed. As we noted above, Kronecker (see Chapter 5) and Hensel (Chapter 6) would later echo this point.

---

[1] See **[39]** for a translation (mostly) of the first edition; see **[61]** for more information on the book and its several editions; see **[19]** for Dedekind's explanation of the supplements.

3.1. Dedekind's Factorization Theory

Since Dedekind's ideal theory has become one of the standard approaches to algebraic number theory, the main features of the factorization theory presented in Supplement X will seem familiar. Dedekind works inside what he calls a "finite field," meaning a subfield of $\mathbb{C}$ that is finite-dimensional over $\mathbb{Q}$. Inside that field he has the ring[2] of algebraic integers. The goal is to create a factorization theory for such integers.

Dedekind follows Kummer in describing the failure of unique factorization: there are irreducible numbers that do not behave as primes should (see Section 1.8). Instead of Kummer's ideal divisors, however, Dedekind considers the *set*[3] of algebraic integers divisible by such a divisor. These sets can be characterized abstractly: they are subsets of the ring of algebraic integers that are closed under sums and under scaling by arbitrary elements. Dedekind called them *ideals*.

An algebraic integer α determines an ideal containing all of its multiples; such ideals are the *principal ideals*. Dedekind defines *prime ideals* and shows that all ideals can be uniquely represented as a product of prime ideals. Thus, while unique factorization fails for the numbers themselves, the corresponding principal ideals can be factored uniquely into prime ideals.[4]

As in Kummer, there is a notion of equivalence that allows us to divide the ideals into finitely many ideal classes. The product of ideals provides a composition of ideal classes. Dedekind goes on to obtain an analytic formula for the number of classes.

All of this is summarized in the Anzeige translated below. Except for the digression into higher congruences Dedekind's account is quite brief.

3.2. Help for Reading Dedekind

The modern reader will find easiest in Dedekind exactly what a contemporary reader likely found hardest: the new abstract notions. There are, however, some points where Dedekind's language is different. There are also parts of the modern conceptual structure that were not yet in place at the time.

On the language side, note that Dedekind's "fields" are always subfields of the complex numbers. He calls a field "finite" if it has finitely many subfields, which is equivalent to having finite degree over $\mathbb{Q}$. Dedekind expresses the fact that one field is contained in another in the language of divisibility: a field "divides" another if it is contained in it.

Similarly, Dedekind's "modules" are always $\mathbb{Z}$-modules contained in $\mathbb{C}$. The name comes from the fact that a module is something that can be used to define congruences.

Ideals are, of course, the key new idea and the main players. Their name comes from Kummer's "ideal primes." Dedekind was unsatisfied with Kummer's conception of an ideal prime, so he replaced it (a typical move for him; see [**67**]) with a set: the set of all numbers divisible by the ideal prime. This works because he is able to characterize these sets abstractly, to define the product of ideals, and

[2]Not Dedekind's term. He calls the collection of all algebraic integers $\mathcal{O}$, but does not have a general name for that kind of algebraic structure.

[3]Dedekind's word is "system."

[4]This shows that $\mathcal{O}$ is a ring with a divisor theory, in the language of [**12**, Section 3.3].

to give a direct definition of what it means for an ideal to be prime. In effect, he replaces "ideal primes" by "prime ideals."

Two important modern concepts were not yet available, making for places where today's reader might have some difficulty. First, Dedekind does not have (or want) the quotient construction. When we would speak of $\mathscr{O}/\mathfrak{p}$ he must talk about congruence classes modulo $\mathfrak{p}$. For example, the (absolute) norm of an ideal is defined as the number of congruence classes modulo that ideal. Second, as we noted above in Section 1.3 of Chapter 1, Dedekind does not use (what we call) finite fields.[5] Instead, Dedekind relies on his paper **[18]**, in which he discusses "double congruences" between polynomials with integer coefficients with respect to a double modulus: a prime p and a polynomial P, usually assumed to have integer coefficients and to be irreducible modulo p. When Dedekind needs to consider $\mathscr{O}/\mathfrak{p}$ as a finite field, he must find a way to translate congruence classes mod $\mathfrak{p}$ to such double congruences. This can be seen, for example, in the first section of Dedekind's 1878 paper discussed in Chapter 4 (see footnote 38 in that chapter). Hensel considered the same issue in 1887; see **[75]**.

Our translation preserves Dedekind's language, but modern equivalents are often given in the footnotes. Our notes also point out links to the material in our other chapters and, occasionally, call attention to things Dedekind does not discuss.

3.3. Translation

Vorlesungen über Zahlentheorie, by P. G. Lejeune Dirichlet. Edited and with additions provided by R. Dedekind. Revised and enlarged second edition. Braunschweig, Friedr. Vieweg und Sohn. 1871.
(Notice by Richard Dedekind)

I discussed the 1863 first edition of this work in these pages (27 January 1864), and so I can refer to that previous article with regard to the origins and content of the book. The new edition differs from the first by way of a great many additions, either in footnotes or to the text itself. Many paragraphs[6] have been completely reworked. These changes, which do not touch the essence of Dirichlet's lectures, mainly reflect the decision to treat, in a new appendix, the tenth supplement, the theory of the composition of binary quadratic forms, which was omitted from the first edition for reasons discussed at that time.[7] The generality with which Gauss presented the theory in the fifth section of the *Disquisitiones Arithmeticae* understandably causes significant difficulties for the beginner. This difficulty motivated Dirichlet to publish *De formarum binariarum secundi gradus compositione*, 1851.[8] He says in his introduction:

[5]As mentioned above, he could not have done so without treating congruence classes as primary objects.

[6]The book, including the supplements, is divided into 170 numbered "paragraphs," which are actually sections, often several pages long. We will generally translate "section" from now on.

[7]From here on everything in the book notice focuses on the content of Supplement X of the second edition **[21]**.

[8]This is **[33]**. See **[111**, 13.7**]** for more on this paper.

> De formarum compositione tunc non egi, quod argumentum ab illustrissimo Gauss in "Disquisitionum Artihmeticarum" sectione quinta maxima quidem generalitate sed per calculos tam prolixos tractatum esse constat, ut perpauci compositionis naturam percipere valuerint, eo magis quod summus geometra, ut ipse monuit, brevitati consulens theorematum difficuliorum demonstrationes synthetice adornavit, suppressa analysi per quam erant eruta. Quare confidere posse mihi videor, hujus argumenti expositionem novam et plane elementarem artis analyticae cultoribus non fore ingratam.[9]

Since in this treatise only the first main theorem of the theory in question is proved and no indication is given of how to continue, I have taken a somewhat different route, which agrees with that of Dirichlet in that only a special case of composition is considered. Sections 145–149 contain the general theorems about the composition of forms and classes of forms.[10] These are used in sections 150, 151 to find the ratio of the class numbers of two determinants whose ratio is a square; this is the same problem treated according to Dirichlet's method in sections 97, 99, 100.[11] There follow in sections 152–154 the composition of genera and Gauss's second proof of the quadratic reciprocity theorem. Sections 155–158 contain a proof of Gauss's theorem that the duplication of any class results in the principal genus; it is based on a theorem of Lagrange and Legendre about the rational integer solutions of indeterminate equations of degree two in two unknowns.

In the paragraphs that follow I tried to introduce the reader to a higher domain[12] in which algebra and number theory are intimately connected. In the course of my lectures on circle division and higher algebra, held in Göttingen in Winter[13] 1856–1857 before Mr. Sommer and Mr. Bachmann and in Winter 1857–1858 before Mr. Selling and Mr. Auwers, I was convinced that the study of the algebraic properties of numbers is most appropriately based on concepts that are directly linked to the simplest arithmetic principles.[14] I replaced the name "rational domain" with the

---

[9]The quote, in Latin, is not set out in the original. An idiomatic English translation might be: "I did not then [in **[32]**] take up the composition of forms, which topic was treated by the most illustrious Gauss in the fifth and largest section of this *Disquisitiones Arithmeticae*. That treatment is surely very general, but it requires such long calculations that very few are able to grasp the nature of the composition. This is all the more so because the great geometer, who himself recommended keeping an eye to brevity, gave demonstrations of the more difficult theorems by synthesis, with the analysis by which they were unearthed suppressed. For that reason it seems I may be confident that a new and plainly elementary exposition of this theory will not be unwelcome to the cultivators of the analytic art."

[10]Since these are standard topics, Dedekind does no more than to say that they are in Supplement X by mentioning them by name.

[11]Dirichlet, following Gauss, assumes quadratic forms look like $ax^2 + 2bxy + cy^2$. Forms like $x^2 + xy + y^2$ are replaced by $2x^2 + 2xy + 2y^2$. As a result, they must allow forms where $\gcd(a, 2b, c) \neq \gcd(a, b, c)$. This is the difficulty treated in sections 97–100 of **[35]**.

[12]This is the unannounced leap, the content that is not at all suggested by the title of Supplement X. To treat binary quadratic forms Dedekind wants to consider quadratic number fields. Since it is no harder (!) to treat all number fields, he proceeds to do so.

[13]The Winter Term in German universities ran roughly from October to February.

[14]See **[73]** and **[43]** for reconstructions of the development of Dedekind's theory. See **[68]** for an analysis of what Dedekind meant by "simplest arithmetic principles."

word "field,"[15] by which I understand a system of infinitely many numbers[16] that has the property that the sums, differences, products, and quotients of two such numbers belong to the same system. I say a field[17] Q is a divisor of a field M, and the latter a multiple of the first, when all the numbers contained in Q are also found in M.[18] Any two fields A, B always have a least common multiple, which can be denoted by AB, and also a greatest common divisor.[19] When to each number a in a field A there corresponds a number $b = \varphi(a)$ so that $\varphi(a+a') = \varphi(a) + \varphi(a')$ and $\varphi(aa') = \varphi(a)\varphi(a')$, the the numbers b make up a field $B = \varphi(A)$ that is conjugate to A, which arises from A by the substitution[20] φ. These concepts are connected, in the algebraic direction, with the ideas of Galois and, in the number-theoretical side, with Kummer's creation of the ideal numbers.[21]

In Section 159 are developed the general properties of a field Ω that has only a finite number of divisors.[22] In such a field there are always finitely many[23] numbers $\omega_1, \omega_2, \ldots, \omega_n$ with the property that any given number ω from the field can always and in only one way be expressed as

$$h_1\omega_1 + h_2\omega_2 + \cdots + h_n\omega_n,$$

where $h_1, h_2, \ldots, h_n$ are rational numbers, which are called the coordinates of the number ω with respect to the basis[24] $\omega_1, \omega_2, \ldots, \omega_n$. The number n is called the degree of the field Ω. It follows quite easily that every number in the field is an algebraic number, namely the root of an equation of degree n whose coefficients are rational numbers. There are n different substitutions linking the field Ω to conjugate fields. The product of the n values obtained from a given number ω via these n substitutions is called the norm of ω. It is a homogeneous function of the coordinates with rational coefficients,[25] therefore a rational number, which is denoted by $N(\omega)$. Given a system of n numbers $\alpha_1, \alpha_2, \ldots, \alpha_n$ from the field Ω, one builds the determinant of the n^2 corresponding numbers in the n conjugate

[15] Dedekind's word is of course *Körper*, which translates as "body." The standard English term is "field," which we will use throughout.

[16] By "numbers" Dedekind seems to mean complex numbers. So his fields are all subfields of $\mathbb{C}$. Notice that Dedekind's concept is intrinsically infinitary: a field is a single object that includes infinitely many numbers.

[17] An important move happens here, quietly. As Edwards says, "the idea of having a single, coordinate-free letter such as K to represent the set of all elements of the field under discussion – as opposed to a notation such as $\mathbb{Q}(\alpha)$ that suggests a particular basis for the field – was original with Dedekind and stemmed very naturally from his point of view." [**43**, p. 343n]

[18] So "Q is a divisor of M" means $Q \subset M$.

[19] We would call the lcm the compositum (inside $\mathbb{C}$) and the gcd the intersection of the two fields A and B.

[20] Dedekind uses "substitution" for what we would call a function, here a field homomorphism.

[21] As Dedekind said at the beginning of the paragraph, he is consciously creating a link between algebra and number theory, perhaps inspired by his teacher Dirichlet's linking analysis and number theory.

[22] Rather than use the dimension to define a finite extension, Dedekind focuses on the number of subfields, which allows him to stick to his "sounds like arithmetic" point of view. But he immediately points out that having finitely many subfields implies that there is a finite basis.

[23] The original is "eine endliche Anzhahl von Zahlen," literally "a finite number of numbers." We went for the more standard English "finitely many numbers."

[24] Dedekind does not call attention to the word "basis" here. That term is defined in Supplement X, as are "dependent" and "independent" in this context.

[25] Dedekind remains attentive to the theory of forms, which is the ostensible subject of the supplement; the norm is a form of degree n in n variables. As noted above, both Dirichlet and Hermite had written about such "decomposable forms."

fields. The square of this determinant is a rational number, which I call[26] the discriminant of the numbers α_1, α_2, ..., α_n and denote by $\Delta(\alpha_1, \alpha_2, \ldots, \alpha_n)$. It is not possible and not necessary to go into the analytical developments[27] that are linked to these concepts; they are only given in this paragraph to the extent that it seemed appropriate for a better understanding.

In the following Section 160 all the algebraic numbers (which also form a field) are divided into integral and fractional numbers. An integer[28] is understood to mean a root of an equation with highest coefficient $= 1$ and whose other coefficients are rational integers. From this concept simple propositions about divisibility, units, and relatively prime numbers are derived for later use.

The following Section 161 contains a lemma[29] for our theory allowing a generalization of Gauss's notion of congruence between numbers.[30] By a module I understand a system $\mathfrak{m}$ of numbers[31] whose sums and differences still belong to the same system. The congruence $\omega \equiv \omega' \pmod{\mathfrak{m}}$ means that the difference $\omega - \omega'$ belongs to the system $\mathfrak{m}$. This concept has a broader scope than its extraordinary simplicity seems to promise, but we only give here what will serve to facilitate the subsequent presentation.

After these preparations, the integers of a field Ω of degree n are investigated in Section 162. They form a module $\mathcal{O}$, and it is shown first[32] that one can find n integers ω_1, ω_2, ..., ω_n that are basis numbers of the field, so that any integer can be represented as

$$\omega = h_1\omega_1 + h_2\omega_2 + \cdots + h_n\omega_n,$$

where all the coordinates h_1, h_2, ..., h_n are whole numbers. The discriminant $\Delta(\omega_1, \omega_2, \ldots, \omega_n)$ of such a basis, which I call a fundamental series,[33] has the smallest possible absolute value. This nonzero rational integer is of particular significance

---

[26]This extension of the notion of discriminant of a polynomial seems to have appeared first here and in Supplement X. The connection is as follows. If the field Ω is obtained by adjoining an element ρ that is a root of an irreducible polynomial $f(x)$ of degree n, then the system $1, \rho, \rho^2, \ldots, \rho^{n-1}$ is a basis. The conjugates of ρ are the other roots of $f(x)$, so that we get the Vandermonde determinant whose columns are the powers of the roots; the square of this determinant is the discriminant of $f(x)$. The generalization, then, allows general bases of Ω instead of only power bases.

[27]To a modern reader, the end of Section 159 is very hard to follow; perhaps Dedekind's readers would have agreed, since he skips all of it here.

[28]Dedekind consistently writes "ganze Zahl" for an algebraic integer and "ganze rationale Zahl" for an element of $\mathbb{Z}$. We translate "integer" and "rational integer" respectively.

[29]Hülfssatz.

[30]This typical Dedekindian move feels perfectly comfortable for the modern reader, but it was not the way things were usually done in the 19th century. Dedekind here introduces a new algebraic idea, a "module," and proceeds to establish the basic properties before returning to the theory of fields.

[31]Dedekind's modules are free $\mathbb{Z}$-submodules of $\mathbb{C}$. Their name comes from the fact that it is possible to define congruences modulo a module.

[32]The first step is to show the existence of an integral basis. A few lines later Dedekind will use the term "fundamental series" for such a basis. That is also the term used in the Kronecker school.

[33]Grundreihe.

for the field Ω, it is called the discriminant or the fundamental number[34] and denoted by $\Delta(\Omega)$. It divides the discriminant of any system of n integers, and the quotient is a square. Furthermore, if μ is a nonzero number in $\mathcal{O}$, the number of incongruent integers with respect to μ is equal to the absolute value of the norm $N(\mu)$. We then draw attention to a strange phenomenon,[35] first observed in the case of cyclotomic fields. It consists in this: an integer that cannot be decomposed as a product of other integers does not always play the role of a true prime number. This was the starting point for Kummer's creation of ideal numbers.

My goal in Section 163 is to propose a theory[36] that applies to all [finite] fields. The fundamental idea is as follows. If μ is a nonzero number in $\mathcal{O}$, then the system $\mathfrak{m}$ of all numbers in $\mathcal{O}$ that are divisible by μ has the following two properties:

I. The sum and difference of two numbers in $\mathfrak{m}$ is a number in $\mathfrak{m}$; that is, $\mathfrak{m}$ is a module.

II. Every product of a number in $\mathfrak{m}$ with a number in $\mathcal{O}$ is also a number in $\mathfrak{m}$.

It is not true that conversely, every system $\mathfrak{m}$ of integers from a field that has these two properties, which from now on I will call an ideal, is always the set of numbers that are divisible by some fixed μ. When this is the case, I say $\mathfrak{m}$ is a principal ideal and denote it by the symbol $\mathfrak{i}(\mu)$.[37] We then investigate the properties of all the ideals of the field Ω, and the following main result follows. Multiplying each number of an ideal $\mathfrak{a}$ by each number of an ideal $\mathfrak{b}$, these products and their sums make up an ideal, which is the product of the two factors $\mathfrak{a}$ and $\mathfrak{b}$ and is denoted by $\mathfrak{ab}$.[38] It then clearly follows that $\mathfrak{a}\mathcal{O} = \mathfrak{a}$, $\mathfrak{ab} = \mathfrak{ba}$, $(\mathfrak{ab})\mathfrak{c} = \mathfrak{a}(\mathfrak{bc})$, and that from $\mathfrak{ab} = \mathfrak{ac}$ it follows that $\mathfrak{b} = \mathfrak{c}$. One says an ideal $\mathfrak{p}$ different from $\mathcal{O}$ is a prime ideal when it has no factors different from $\mathcal{O}$ and $\mathfrak{p}$;[39] a composite ideal can be decomposed as a product of prime ideals and in only one way. One then defines the norm $N(\mathfrak{a})$ of an ideal $\mathfrak{a}$ to be the quantity of numbers in $\mathcal{O}$ that are incongruent with respect to the module $\mathfrak{a}$. We have $N(\mathfrak{ab}) = N(\mathfrak{a})N(\mathfrak{b})$. In this way we obtain a complete analogy with the laws of divisibility in rational number theory.[40]

---

[34] Grundzahl. There are at this point three different things called "discriminant": the discriminant of a polynomial (Gauss's "determinant"), the discriminant of a set of n algebraic numbers, defined above, and the discriminant of a field, the Grundzahl. Sylvester's goal (see Section 1.6) has been thoroughly frustrated.

[35] The "strange phenomenon" is the failure of unique factorization, but Dedekind, following Kummer, describes it by saying that some irreducible elements of $\mathcal{O}$ do not behave like true primes.

[36] A theory of factorization is meant. The ideal primes of Kummer will be replaced by prime ideals.

[37] In **[23]**, the notation was changed to either $\mathcal{O}\mu$ or $\mathcal{O}(\mu)$, the latter when μ is an explicit number.

[38] It is often said (e.g., [**113**, 1.1.3]) that this definition of the product of two ideals is not actually in Supplement X. In fact it is, in §163, item 6 on page 459. But it is Dedekind's second definition of the product of two ideals, and Dedekind checks in that section that the two definitions agree. By the time Dedekind wrote the Anzeige, he had clearly become convinced that this was the better definition.

[39] Dedekind's definition of prime ideal sticks to the analogy with ordinary arithmetic. In modern terms it is actually the definition of a *maximal* ideal. This definition is shown to be equivalent to the modern definition in **[23]**; see Section 4.3, §2 on p. 55 below.

[40] This completes the description of Dedekind's theory of ideals and their factorization.

This entire theory is intimately connected with the so-called theory of higher congruences,[41] which was suggested by Gauss and developed in the work of Galois, Schönemann and others.[42] The works of Kummer on cyclotomic ideal numbers and the study of the algebraic investigations of Galois led me to consider the theory of higher congruences, and I published a brief outline of that theory (Crelle's Journal, Vol. 54).[43] I later sought, with its help, to create a general theory of ideal numbers, but was distracted from it by other work until the publication of this[44] work led me back to that subject. The renewed effort led me to my new theory of ideals, which seems preferable to me because it is based on much simpler concepts.[45] In my presentation I did not deal closely with the connection with the theory of higher congruences, because I feared that the extent of my appendix would become too large.[46] For readers who are interested in this connection,[47] I hope the following remarks may be useful.[48]

Let ω be an arbitrary number in $\mathscr{O}$, and set[49]

$$\Delta(1, \omega, \omega^2, \ldots, \omega^{n-1}) = D^2\Delta(\Omega).$$

Then D is always a rational integer, namely a homogeneous function of degree $\frac{1}{2}n(n-1)$ of the coordinates with rational integer coefficients.[50] If then p is a rational prime number and we are given a number ω for which D is not divisible by p, then the decomposition of the principal ideal $\mathfrak{i}(p)$ as a product of prime ideals is easily found via the theory of higher congruences.[51] The number ω satisfies an equation of degree n $F(\omega) = 0$ and if

$$F(x) \equiv P_1(x)^{e_1} P_2(x)^{e_2} \ldots P_m(x)^{e_m} \pmod{p},$$

where P_1, P_2, ..., P_m are pairwise distinct prime functions[52] of the variable x of degrees f_1, f_2, ..., f_m respectively, then we have

$$\mathfrak{i}(p) = \mathfrak{p}_1^{e_1}\mathfrak{p}_2^{e_2} \ldots \mathfrak{p}_m^{e_m},$$

---

[41]Here Dedekind veers off the track. So far he has given a blow-by-blow account of Supplement X, but none of the material on higher congruences is found there.

[42]See Section 1.3 for references to the works Dedekind mentions here.

[43]This is **[18]**, which discusses both congruences between polynomials and "double congruences" that amount to a theory of finite fields, hence the reference to Galois. See the discussion in Section 1.3.

[44]Presumably the publication of the first edition of the *Vorlesungen*? Or the decision to write Supplement X?

[45]Many of Dedekind's contemporaries did not feel Dedekind's approach was in any way "simpler." In particular, there was a lot of resistance to working with infinite sets as objects. As such, ideals seemed very abstract.

[46]The supplement was 118 pages long; in later editions it was split into supplements X and XI and came to be much longer.

[47]By which we suspect he means those who want to know why Dedekind did not stick to the more straightforward congruence-based approach.

[48]Here begins the digression; this material is not in Supplement X.

[49]Dedekind does not mention that this is the discriminant of the polynomial of degree n with ω as a root, but of course that was the original sense of "discriminant" that was generalized to n-tuples just above. He also does not mention the possibility that $D = 0$; he is more explicit about this in **[23]**.

[50]This homogeneous function was later called the "index form."

[51]This is called "Dedekind's Theorem" in many modern textbooks. The proof was first given in **[23]**. See Section 4.3, §2.

[52]Dedekind uses "prime function" for irreducible polynomial. Here he means that they are irreducible modulo p.

where $\mathfrak{p}_1, \mathfrak{p}_2, \ldots, \mathfrak{p}_m$ are pairwise distinct prime ideals with norms $p^{f_1}, p^{f_2}, \ldots p^{f_m}$, respectively. From this the following theorem, which is fruitful for both algebraic and number-theoretic investigations, follows[53] easily:

The prime number p divides the fundamental number $\Delta(\Omega)$ of the field Ω if and only if p is divisible by the square of a prime ideal.[54]

At first I thought it very likely[55] that for any given prime number p there would exist an integer ω such that the number D was not divisible by p.[56] Only when all my attempts to prove the existence of such a number were unfruitful did I set myself the task of investigating whether this conjecture was incorrect.[57] Were the conjecture true, whenever p is divisible by r distinct prime ideals $\mathfrak{p}$ whose norms have value p^f, there must exist r distinct prime functions P of degree f.[58] Conversely, when this last condition is always satisfied, then one can prove the existence of a number ω with the desired property.[59] In the simplest case when $f = 1$, there are exactly p distinct prime functions of degree one.[60] The question then becomes whether there exists a field Ω in which p is divisible by[61] $(p+1)$ distinct prime ideals, all of them of norm p. The degree of such a field must then be $= p+1$. The simplest case arises when one takes $p = 2$, leading to the question: do there exist cubic fields in which the number 2 is divisible by three distinct prime ideals? In such a field D would always be an even number.[62] One can always

---

[53]As Dedekind will clarify, it follows "easily" only for primes that do not divide D: if p divides $\Delta(\Omega)$, then $F(x)$ has multiple roots modulo p, hence is divisible by the square of an irreducible polynomial. Assuming p does not divide the index, it follows from the factorization theorem that it is divisible by the square of a prime ideal. The theme is picked up at the end of this section (page 41) and again at the end of Section 4.3 (page 74).

[54]No italics in the original, but the statement does get its own paragraph. The proof that this is true for all primes first appeared in 1882 **[24]**, where Dedekind introduces the notion of the *different* to obtain the proof.

[55]See the discussion in Section 2.4, where we also note the very similar comments in **[23]**, translated below; see opening paragraph of Section 4.3, §4 on page 63.

[56]One reason to hope for this is that it would provide a way of finding the factorization in $\mathcal{O}$ of any rational prime p.

[57]As indeed it is, which Dedekind will show. Kronecker claimed that he knew an example of this in 1858. See Section 2.5 and the latter part of Section 5.4 on page 86.

[58]In effect, Dedekind is telling us that his factorization theorem implies that if p is not an index divisor then there must exist sufficiently many irreducible polynomials of the given degrees in $\mathbb{F}_p[x]$. Contrapositively, if there are not enough irreducible polynomials, then p is an index divisor. The theorem that motivated the search for a good generator ω already implies that one does not always exist.

[59]Dedekind also claims that his condition is sufficient: if there are enough irreducible polynomials, then one can find a generator for which p is not an index divisor. This is one of the main results Dedekind proved in **[23]**; it is proved again in **[78]**. See Section 4.3, §4, and Section 6.4, §1.

[60]Namely, $x - a$, $a \in \mathbb{F}_p$.

[61]Dedekind should have said "at least $p + 1$," but he is looking for the simplest possible example.

[62]Dedekind has twice reduced to the "simplest case" in order to find his example: first he assumes $f = 1$ and then $p = 2$. The task now is to find a cubic field in which 2 splits completely. By Dedekind's theorem this would require three different irreducible polynomials of degree 1 in $\mathbb{F}_2[x]$. Since there are only two such, the theorem cannot hold, and so D must be even for every choice of ω. So the question becomes: can we find a cubic field in which the prime 2 splits completely? Dedekind did not have any means of answering this directly, so instead he constructs something like the *generic* cubic field then sets the parameters so that D must be even. From that he concludes that 2 splits completely. One result of this strategy is that he can show 2 is CIDD *without* relying on the factorization theorem, whose proof would only appear in 1878.

assume that the fundamental series[63] of a cubic field consists of the number 1 and two integers α, β whose product is rational.[64] One then has[65]

$$\begin{aligned}\alpha\alpha &= a'\alpha + b\beta - bb'\\ \beta\beta &= a\alpha + b'\beta - aa'\\ \alpha\beta &= ab\end{aligned}$$

where a, b, a', b' are rational integers with no common divisor,[66] and we can compute[67]

$$\begin{aligned}\Delta(\Omega) &= \Delta(1,\alpha,\beta)\\ &= a'^2b'^2 + 18aba'b' - 4aa'^3 - 4bb'^3 - 27a^2b^2.\end{aligned}$$

If we now set

$$\omega = z + x\alpha + y\beta,$$

with z, x, y any rational integers,[68] then

$$\omega^2 = z^2 - bb'x^2 - aa'y^2 - aa'y^2 + 2abxy \\ + (a'x^2 + ay^2 + 2xz)\alpha + (bx^2 + b'y^2 + 2yz)\beta,$$

and it follows[69] that

$$D = bx^3 - a'x^2y + b'xy^2 - ay^3,$$

independent of z, which is expected[70] from the definition of D.[71] Even though a, b, a', b' have no common divisor, D will be an even number whenever a and b are

[63]Since integral bases exist, the ring of integers of a cubic field Ω is a free $\mathbb{Z}$-algebra of rank three, what we might call a "cubic algebra." Dedekind takes the basis to be $\{1, \alpha, \beta\}$ and considers the corresponding multiplication table, simplifying as much as possible. This provides a number of parameters that can be adjusted to obtain the desired property. Compare this to **[23]**, Section 4.3, §5 on page 68 below, where he starts from a polynomial equation defining the field.

[64]If $\alpha\beta = \ell\alpha + m\beta + n$, replace α by $\alpha - m$ and β by $\beta - \ell$.

[65]Dedekind offers no explanation for why the formulas should look like this. Define integers a, a', b, b', c, c', n by $\alpha\beta = n$, $\alpha^2 = a'\alpha + b\beta - c$ and $\beta^2 = a\alpha + b'\beta - c'$. Then notice that $n\beta = \alpha\beta^2$. Expanding the latter and equating basis coefficients gives $n = ab$, $c = bb'$, $c' = aa'$, as Dedekind says. The only assumption needed is that we are working in a free $\mathbb{Z}$-algebra of rank three.

[66]The fact that the four parameters do not have a common divisor is not actually used in what follows, because Dedekind will choose their values. But if we know that our $\mathbb{Z}$-algebra has a divisor theory, which of course is true for $\mathcal{O}$, then the parameters must have no common divisors. The key observation is that in a ring with a divisor theory $\alpha^2|\beta^2$ implies $\alpha|\beta$. If some rational prime divides all four integers, then p^2 would divide α^2, $\alpha\beta$, and β^2, and hence p^2 would divide $(\alpha+\beta)^2$, and so $\frac{1}{p}\alpha + \frac{1}{p}\beta \in \mathcal{O}$, contradicting the assumption that $\{1, \alpha, \beta\}$ is an integral basis.

[67]Given the information we have, we can compute the traces of α^2, β^2, and $\alpha\beta$; from that information it is easy to compute the discriminant.

[68]This is surprisingly close to Kronecker's "generic element" of $\mathcal{O}$; see Chapter 5. In the notation of that chapter, $\Delta(\Omega)$ is Kronecker's D and Dedekind's $D^2\Delta(\Omega)$ becomes Kronecker's $\mathcal{D}$ if we take x, y, z as variables.

[69]We have expressed each element of the basis $\{1, \omega, \omega^2\}$ as a linear combination of $\{1, \alpha, \beta\}$; D is the determinant of the base-change matrix.

[70]If z is an integer the discriminants of $f(t)$ and of $f(t+z)$ are the same.

[71]Dedekind has not put any restrictions on the values of x, y, z. Bad choices (most obviously, $x = y = 0$) will result in $\{1, \omega, \omega^2\}$ not being a basis, which would make $D = 0$.

even and a' and b' are odd.[72] It must then be[73] that the number 2 is divisible by three distinct prime ideals. This is completely confirmed by the example[74]

$$a = b = 2, \quad a' = -b' = 1, \quad \Delta(\Omega) = -503;$$

we have[75]

$$\mathfrak{i}(2) = \mathfrak{abc}, \quad \mathfrak{i}(\alpha) = \mathfrak{a}^2\mathfrak{c}, \quad \mathfrak{i}(\beta) = \mathfrak{b}^2\mathfrak{c},$$

where $\mathfrak{a}$, $\mathfrak{b}$, $\mathfrak{c}$ are three distinct prime ideals.[76]

Another example can be obtained in the following way. With respect to the modulus $p = 2$ there exists only one prime function of degree two, namely x^2+x+1. Therefore when in a field Ω the integer 2 is divisible by at least two distinct prime ideals whose norm $= p^2 = 4$, then D must be even. In this case the degree of the field must be at least $= 4$. The phenomenon in fact occurs in the biquadratic field[77] defined by the equation

$$\alpha^4 - \alpha^3 + \alpha^2 - 2\alpha + 4 = 0.$$

The numbers 1, α, $\beta = 2 : \alpha$, and $\gamma = \alpha^2 - \alpha$ are a fundamental series and the fundamental number is $= 13^2 \cdot 17$.

Thus, there exist fields Ω in which the number D above is always divisible by certain singular prime numbers p. Of course there are only finitely many such primes. I remark, however, that the theorem[78] above, characterizing of the rational primes that divide the fundamental number $\Delta(\Omega)$ of a field, remains valid in general, but it would take us too far afield were I to give a proof of this theorem or to explore its significance for the theory of fields.

After this digression, I continue to summarize the contents of the sections that follow.[79] In Section 164 the ideals of the field Ω are divided into a finite number of classes. Two ideals are called equivalent when their product by some fixed ideal

---

[72]If a, b are even and a', b' are odd,

$$D \equiv x^2y + xy^2 \equiv xy(x+y) \equiv 0 \pmod 2$$

for all integers x, y. Thus, Dedekind has shown directly (and independently of the factorization theorem) that 2 is a common index divisor for any cubic algebra of this form.

[73]Dedekind does not say why, but this follows from the characterization of common inessential discriminant divisors that he gives in his 1878 paper translated in the next chapter. He has mentioned the key result above: when there are enough polynomials one can find a good generator.

[74]At no point has Dedekind checked that his $\mathbb{Z}$-algebra is actually a domain, so that Ω is a field. His choice $(a, b, a', b') = (2, 2, 1, -1)$ suggests, however, that he did notice that this is not automatic. To see this, note that the minimal polynomial for α is $x^3 - a'x^2 + bb'x - ab^2$. While this is irreducible for most choices of the quadruple (a, b, a', b'), that is not always the case. Dedekind's choice gives $x^3 - x^2 - 2x - 8$, which is indeed irreducible and so we have a cubic field of discriminant -503, global number field 3.1.503.1 in the LMFDB database [**109**]. But if we chose $(2, 2, 1, 1)$ we would get $x^3 - x^2 + 2x - 8 = (x-2)(x^2+x+4)$ (any choice with $a = b$ and $a' = b'$ gives a reducible polynomial). Even more dramatically, the quadruple $(6, 2, 9, 13)$ would give $\Delta = 1$, which is impossible for a number field; it corresponds to the polynomial $x^3 - 9x^2 + 26x - 24 = (x-2)(x-3)(x-4)$.

[75]To confirm that he has the example he wants, Dedekind writes out (without proof) the factorizations of 2, α, and β.

[76]See the more complete discussion in [**23**], translated in Section 4.3, §5, where Dedekind defines the three ideals explicitly and checks all of these statements.

[77]In this case Dedekind does not tell us how he found the field; presumably the method was similar. This is global number field 4.0.2873.1 in the LMFDB database [**109**].

[78]That is, the theorem that the primes dividing the fundamental number are exactly those divisible by the square of a prime ideal.

[79]We are back to Supplement X.

is a principal ideal. An ideal class consists of all ideals that are equivalent to a given ideal. The principal class consists of the principal ideals. These ideal classes allow a composition that has the same properties of the composition of classes of quadratic forms.

In Section 165 I show the relationship between the composition of ideal classes and the decomposable homogeneous forms that arise from the same field Ω.

Section 166 gives Dirichlet's theory of units in a generalized form.[80] The presentation is completely independent of the previous one. In Section 167 this theory is used to obtain an expression for the number of ideal classes by way of an infinite series, much like the determination of the class number of quadratic forms. At this point I break away from the study of the general problem, since my investigations of this topic have not yet been crowned with sufficient success to be published.[81] The sections that follow, 168–170, illustrate the general theory by applying it to the example of quadratic fields.

So far it appears that the theory of ideal numbers has been the subject of serious research by only four or five mathematicians.[82] My heartfelt wish is that the new edition of Dirichlet's *Vorlesungen über Zahlentheorie* may facilitate access to this large subject and perhaps to motivate a larger number[83] of mathematicians to apply their powers so that, amidst the tremendous recent progress made in geometry and in the theory of functions, number theory may not be left behind.

July 22 1871 R. Dedekind

3.4. Dedekind's Strategy in his *Anzeige*

In both the original version of Supplement X and the book notice calling attention to it, Dedekind opts for a kind of nested structure. What seems fairly straightforward from the title and the first few paragraphs hides a surprise in the middle. In Supplement X, apparently dealing with the theory of composition of binary quadratic forms, the surprise is a general theory of divisibility in algebraic number fields. In the book notice, the surprise is the inclusion of a discussion of higher congruences and common inessential discriminant divisors.

It was a strange strategy. In 1876, Dedekind would tell Rudolf Lipschitz that he "thought that the inclusion of this investigation in Dirichlet's *Zahlentheorie* would be the best way to attract a wider circle of mathematicians to the field,"[84] but it seems unlikely that many mathematicians would look for an entirely new theory in the middle of an appendix to a textbook, an appendix ostensibly treating material they already knew.

---

[80]Dirichlet always worked in $\mathbb{Z}[\alpha]$ for some algebraic integer α. In Supplement X, Dedekind gave the proof for the full ring of integers. The argument is essentially the same.

[81]As far as we know, Dedekind never returned to the theory of composition of decomposable n-ary forms of degree n.

[82]Who were they? We think the likely names are Kummer, Kronecker, Bachmann, and Selling.

[83]Famously, this did not happen. See our discussion in Section 3.4.

[84]"Ich hatte geglaubt, dass die Aufnahme dieser Untersuchung in Dirichlets Zahlentheorie das sicherste Mittel wäre, um einen grösseren Kreis von Mathematikern für die Bearbeitung dieses Feldes zu gewinnen..." Translation by John Stillwell in [**26**, Section 0.7.1]. For the original, see [**108**, p. 49].

As it happens, we know the strategy failed. The correspondence in 1876 was prompted by a letter from Lipschitz, who said that he thought Dedekind's investigations are "of rare value" but had not received the attention they deserved, "not even in Germany."[85] In his reply, Dedekind admits that the only person who had expressed any interest in ideal theory was Heinrich Weber, Dedekind's collaborator in the edition of Riemann's collected papers. This led to an extended correspondence between Dedekind and Lipschitz (see pages 47–106 in [**108**]) and also to the publication in French of a second version of Dedekind's theory (see [**22**]; for an English translation, see [**26**]).

It seems likely that the discussion of higher congruences and common index divisors was included in the Anzeige in order to answer a question Dedekind anticipated: why not do it the "natural" way? Everyone who had attempted to follow Kummer's lead had introduced ideal primes by way of congruences. Dedekind's approach would have seemed unexpected and strange. Many found it so. Zolotarev, for example, notes the work of Selling and of Dedekind, but says "If I am not mistaken, so far there is no theory of complex numbers for the case of general equations that is as satisfying as M. Kummer's theory for binomial equations."[86] Dedekind must have hoped that including a discussion of the problems facing an approach through higher congruences would help motivate his decision to do something different.

The material on higher congruences in Dedekind's *Anzeige* contains a number of key insights. It was easy to see in previous work that one could use higher congruences to determine the factorization of a rational prime p as long as p was not a divisor of the discriminant of the polynomial that defines the field $\Omega = \mathbb{Q}(\theta)$. Dedekind's first key insight was that the real problem was primes dividing the *index* of $\mathbb{Z}[\theta]$ in $\mathcal{O}$ rather than all the divisors of the polynomial discriminant. The obstruction lay in Kronecker's "inessential divisors." The second key insight was that because the number of irreducible polynomials of a given degree in $\mathbb{F}_p[x]$ is *finite*, the factorization theorem itself already implied that in some cases the index would *always* be divisible by certain primes. This destroyed any hope that a complete theory of ideal primes could be constructed simply on the base of "higher congruences" and led Dedekind to look for an alternative approach.

Dedekind's approach to constructing an explicit example of a field with common inessential discriminant divisors is particularly interesting. At the time (and, we think, still today), most mathematicians approached algebraic numbers via the polynomial equations that defined them. When seeking to construct a field Ω, the straightforward approach was to look for an irreducible polynomial that defined it. That is what Dedekind did when he discussed the same example in his 1878 paper.

That, however, is not at all what Dedekind did in the *Anzeige*, perhaps because he wanted to explain how he had constructed his example. He started by writing down an integral basis for the ring of integers he was trying to construct. Using the properties he expected his ring to have, he showed that any such ring could be determined by a choice of four integer parameters (a, b, a', b'). Choosing the parameters appropriately would produce the field he wanted. Indeed, varying the

[85]"... meiner Meinung nach von einem zeltenen Werthe sind, und dabei selbst in Deutschland nicht nach ihrer vollen Gebühr allgemein gewürdigt werden..." [**108**, p. 47]

[86]"Mais, si je ne me trompe, jusqu'ici il n'y a pas de théorie des nombres complexes pur le cas des équations quelconques aussi satisfaisante que la théorie de M. Kummer pour le cas des équations binômes." [**144**, p. 52]

choice of parameters we can produce infinitely many examples of cubic fields in which 2 is a CIDD. Since Dedekind also gave a quartic example, he may have done something similar for $\mathbb{Z}$-algebras of rank four.

Did Dedekind notice that his construction would in some cases yield a $\mathbb{Z}$-algebra that was not the ring of integers of some field? It seems that he must have, since he avoided the most obvious case in which that happens. He does not, however, mention the issue at all.

CHAPTER 4

Dedekind on Higher Congruences

By 1878, Dedekind had given two accounts of his theory of ideals. The first is [**21**], Supplement X of the second edition of Dirichlet's *Vorlesungen über Zahlentheorie* and discussed in the "Anzeige" translated in the previous chapter. The second is an article [**22**] published in French in several parts, 1876–77. In the paper we translate and analyze in this chapter, Dedekind refers to these as "D" and "B"; for an English reader, B is the preferred reference, since it was translated by John Stillwell and published by Cambridge University Press [**26**]. A third edition of the *Vorlesungen* appeared in 1879; it contained a rewritten version of the theory of ideals that is quite close to the version in B. A fourth edition would follow, with still another version of the theory.

Shortly after [**22**] was published, Dedekind tells us, he became aware of a paper by Zolotarev that attempted to construct a general theory of algebraic integers on the basis of polynomial congruences. Since Dedekind could not read Russian, all he had was a short abstract of the paper. That seems to have motivated him to write the paper [**23**], "Über den Zusammenhang zwischen der Theorie der Ideale und der Theorie der höheren Kongruenzen," which we translate below. The title of translates as "On the Relationship between the Theory of Ideals and the Theory of Higher Congruences." It tells us that the paper will discuss the connections between two subjects about which Dedekind had already written in 1871: the theory of "higher congruences" and the theory of ideals. In fact, this paper contains proofs of several of the results announced in 1871.

In this chapter we give an outline and an annotated translation of [**23**]. We have tried to translate the mathematical content precisely, usually preserving Dedekind's terminology. Our annotations are given as numbered footnotes; Dedekind's own footnotes are marked with non-numerical footnote markers.

When the paper was republished in the first volume of Dedekind's collected works [**25**], Öystein Ore added several endnotes, which we give in summary form at the end. The editors also added a few footnotes that we have translated in annotations, distinguishing them from Dedekind's original footnotes. There are several spelling changes made in [**25**]; for example, "Discriminante" becomes "Diskriminante." When we quote the German, we have tried to stick to the original spelling.

Before considering Dedekind's paper, we give a brief account of Zolotarev's work and what Dedekind could have known about it. We then give an outline of Dedekind's paper, which is followed by the actual text and comments.

4.1. Zolotarev's Work

The non-reception of the work of Egor Ivanovich Zolotarev by the German mathematicians working on algebraic numbers and factorization has often been lamented. Not only did he obtain a correct alternative divisibility theory for algebraic integers, he also anticipated methods that later came to be important. By the time his work became available, however, most minds were already made up. Zolotarev's early death also meant that his approach had no champion.

Brief accounts of Zolotarev's life can be found in [**30**], [**118**], and [**51**]. Born in 1847, he was only 31 when he died after a train accident. He and Aleksandr Nikolaevich Korkin worked together on several papers. In the early 1870s, while attempting to generalize a theorem of Chebyshev about integration in finite terms, Zolotarev realized that he needed a theory of ideal divisors for general algebraic numbers. This theory, with the application to integration, became the topic of his 1874 dissertation [**142**].[1] In the dissertation Zolotarev excluded from consideration a finite number of primes. Two papers followed in which he showed how to handle the "exceptional"[2] primes: a partial account published in Russia in 1878 [**143**], and a full version [**144**] that appeared in Liouville's Journal in 1880.[3] Neither Dedekind nor Kronecker could read Russian, but the two papers, written in French, would have been accessible once they were published. The 1878 paper gave no hint that the results it contained were the basis for a theory of ideal divisors. Kronecker mentions [**144**] (see prefatory remarks in Section 5.4), but we do not know whether he or Dedekind ever read it. As Delone points out,[4] however, the first pages of this long paper "reproduce almost word for word Chapter III of his doctoral dissertation." It is quite possible that many readers realized this and never got to the new material that begins in Section 22 of the paper.

Several authors have investigated Zolotarev's work. In [**30**] we find a summary of the dissertation, followed by an account of the two later papers. Delone's account stays quite close to the original notation and terminology. A more modern interpretation appears in [**7**]. It was developed in more detail by Paola Piazza in her thesis[5], summarized in the papers [**118**] and [**119**]. In both [**118**] and [**7**], the emphasis is on how Zolotarev's approach anticipates the notion of localization at a prime p.

Zolotarev's role in the story of CIDDs, as made clear in Dedekind's introduction to the paper below, was mediated by an abstract of his dissertation [**142**] in the *Jahrbuch über die Fortschritte der Mathematik* (Vol. 6, p. 117).[6] This gives more detail about the application to integration than it does to ideal divisors, but it makes clear that in that work Zolotarev was working only with primes that were not index divisors.[7] After noting this restriction, the authors of the abstract say

[1]Zolotarev told Korkin in a letter that he considered his divisor theory more important than the application; see [**30**, p. 66], which unfortunately does not give a date.

[2]The word in [**143**] is "singulaire." In [**144**] he uses "exceptionnel."

[3]As Piazza notes in [**118**], the paper was submitted in 1876. The reason for the long delay in publication is unclear.

[4]See [**30**, p. 77].

[5]Unfortunately, we have been unable to obtain a copy of Piazza's thesis.

[6]The abstract can be accessed at the ZBMath site, `https://zbmath.org/06.0117.02`. The authors are identified as "Ke. (Wn.)", which means "Korkine, Prof. (Petersburg) (Wangerin, Prof. (Berlin))." Thus, Zolotarev's friend and collaborator A. N. Korkin was the main author of the abstract.

[7]Not his term. He used the criterion proved by Dedekind in §3 of the paper below to exclude what he called the "exceptional" primes.

> If this condition is not fulfilled, for a given modulus p one can transform the equation $F(x) = 0$ in such a way that the assumption is fulfilled. The author reserves the discussion of this transformation for another occasion.[8]

Not having seen what Zolotarev actually did in the exceptional case, Dedekind seems to have assumed that the claim was that it was possible to choose a new generator for which p is not an index divisor. Dedekind had shown in 1871 that this cannot always be done. This claim seems to have been his main motivation to write the paper translated below and in particular to give proofs of the theorems announced in 1871.

4.2. Outline

As noted above, Dedekind uses "D" to refer to **[21]**, Supplement X of the second edition of Dirichlet's *Vorlesungen über Zahlentheorie.* He uses "B" for **[22]** or **[26]**. Since the paper deals with higher congruences, Dedekind also frequently refers to his *Abriß* **[18]** of 1857, which includes much of what we would now describe as the theory of finite fields. Dedekind refers to it as "C." There is no "A."

The introduction explains how Dedekind read about Zolotarev's paper and decided to write this one to clarify the relationship between ideals and higher congruences and to provide the proofs of the theorems he had announced in 1871. Dedekind divides the remainder his paper into sections numbered §1 to §5. We summarize the main results in each section.

§1 is all setup. Given a number field $\Omega = \mathbb{Q}(\theta)$, where θ is an algebraic integer, let $\mathscr{O}$ be its ring of integers. Dedekind defines primes of the first kind as those who do not divide the index $(\mathscr{O} : \mathbb{Z}[\theta])$; primes of the second kind are those that do divide the index. Of course, those two "kinds" depend on the chosen generator θ.

§2 contains the proof of "Dedekind's Theorem": for primes p of the first kind, one can determine the factorization of $\mathscr{O}p$ by factoring the minimal polynomial of θ modulo p.

§3 shows that given θ one can decide whether p is of the first or second kind directly from the factorization modulo p of the minimal polynomial of θ and without actually computing the field discriminant.[9] This is important from a practical point of view, since computing the field discriminant can be difficult.

§4 contains the characterization of primes that divide all the element indices, i.e., the common index divisors (Kronecker's common inessential discriminant divisors). Given the degrees $f_1, f_2, \dots, f_m$ of the distinct prime divisors of $\mathscr{O}p$, Dedekind shows that p is a common index divisor if and only there do not exist distinct irreducible polynomials of degrees $f_1, f_2, \dots, f_m$ in $\mathbb{F}_p[x]$. The "if" part is a direct consequence of the theorem in §2, so the main work here is proving the "only if" part. The upshot is common index divisors are a "small primes" phenomenon: they occur exactly when there are not sufficiently many irreducible polynomials in $\mathbb{F}_p[x]$.

§5 presents the simplest example of a field where the criterion in §4 is satisfied. This is the same example as in the "Anzeige" translated in Chapter 3, but the presentation is different, beginning with the polynomial that defines the field. There

---

[8] Ist diese Bedingung nicht erfüllt, so kann man für einen gegebenen Modul p die Gleichung $F(x) = 0$ derart transformiren, dass jene Annahme erfüllt ist. Die Auseinandersetzung jener Transformation behält sich der Verfasser für eine andere Gelegenheit vor.

[9] This characterization was the one used by Zolotarev to define his "exceptinal" primes.

is more detail in the computations in the field but much less information on how the example was found.

4.3. Translation

On the Relationship between the Theory of Ideals
and the Theory of Higher Congruences

by R. Dedekind

The new principles by which I arrived at a theory of ideals that is rigorous and without exceptions were first explained seven years ago in the second edition of the *Lectures on Number Theory* by Dirichlet (§ 159–170) and more recently given, in greater detail and in slightly modified form, in the *Bulletin des sciences mathématiques et astronomiques* (t. XI. p. 278; t. I (2e. serie), p. 17, 69, 144, 207).[10] Stimulated by the great discovery of Kummer, I had been concerned with this subject for many years, starting from a completely different basis, namely the theory of higher congruences. Although these investigations brought me very close to the desired goal, I decided not to publish them, because the theory that emerges suffers from two imperfections. The first is that the investigation of a domain of integral algebraic numbers begins first with the consideration of a certain number and the equation corresponding to it, which is then interpreted as a congruence. The definitions of ideal numbers (or rather of divisibility by ideal numbers) are obtained in this way. Since everything depends on a specific representation, it follows that the invariant character of the definition cannot be recognized from the start.[11] The second imperfection of this approach is that there are peculiar exceptional cases that require special treatment.[12] My more recent theory, on the other hand, is based exclusively on notions such as *fields*, *integers*, and *ideals*, whose definition does not require any particular form of representation of the numbers, removing the first defect. The power of these extremely simple concepts is shown by the fact that, in proving the general laws of divisibility, a distinction between several cases never occurs again. I have made some remarks about the connection between the two types of justification and stated some theorems without proof in the *Göttingischen gelehrten Anzeigen* of September 20, 1871 (pp. 1488–1492). In particular I have discovered the reason[13] for the existence of the peculiar exceptional cases mentioned above. Since then, a theory of ideal numbers by Zolotareff appeared

[10]The first publication is [**21**], 1871, and the second is [**26**], 1876–77.

[11]This is one of Dedekind's fundamental methodological principles: one should always try to define things in a way that is independent of specific choices, rather than making such choices and then proving invariance. He wanted his mathematics "coordinate-free."

[12]Those special cases are the index divisors, one of the main topics of this paper. See also the anecdote recorded in [**25**, vol. 3, p. 481]. Dedekind, unlike Kummer and Zolotarev, does not want to treat the "exceptional" primes differently from the rest.

[13]As we saw in Chapter 3, Dedekind discovered that common index divisors exist if and only if there are insufficiently many irreducible polynomials of the required degrees in $\mathbb{F}_p[x]$. The proof will be given below. Perhaps Kronecker asked Hensel to investigate the problem because he did not think Dedekind's "reason" was a sufficient answer to the question; in any case, Hensel eventually rediscovered the (same) "reason." See Chapter 6.

in 1874, in a paper in Russian with the title *Théorie des nombres entiers complexes, avec une application au calcul integral.*[14] This was announced and abstracted in the *Jahrbuch über die Fortschritte der Mathematik* (Vol. 6, p 117). From the abstract* it is clear that the theory of Zolotareff is also based on the theory of higher congruences, but that the treatment of the aforementioned exceptional cases is temporarily excluded and is reserved for a later presentation. I do not know if this prospective completion[15] has since been published. Since, however, the connection between the two types of justification of general ideal theory is of sufficient interest in itself, I allow myself to provide here the proofs of the remarks given in the *Göttingischen gelehrten Anzeigen.*[16]

I will assume as known both my theory of ideals and the theory of higher congruences, of which I gave a short description earlier in Borchardt's *Journal* (Vol. 54, p. 1).[17] For brevity, I will cite this paper on congruences as C, the second edition of Dirichlet's number theory[18] as D, and the paper in the *Bulletin des sciences mathématiques*[19] as B.

§1[20]

Let Ω be a finite field[21] of degree n, and let $\mathcal{O}$[22] be the domain of all [algebraic] integers[23] contained in it. There always exist n independent integers

$$\omega_1, \omega_2, \ldots, \omega_n$$

which are a basis for the domain $\mathcal{O}$, that is, the system $\mathcal{O}$ is identical with the collection

$$[\omega_1, \omega_2, \ldots, \omega_n]$$

of all numbers ω of the form

$$\omega = h_1\omega_1 + h_2\omega_2 + \cdots + h_n\omega_n.$$

---

[14]Dedkind cites the title in French, following the *JFM*, but the text is in Russian. This is [**142**], which was Zolotarev's final dissertation for a professorship at the University of St. Petersburg; see [**30**, p. 63]

[15]Some elements of the "completion" were sketched in [**143**], which however does not go into the details of the general theory of ideal divisors. The full account of Zolotarev's theory appeared in 1880 as [**144**].

[16]It is clear, then, that Zolotarev's paper was the main stimulus for writing this paper. Dedekind does not prove all the remarks made in 1871 in this paper, however, saving the theorem about ramified primes for [**24**].

[17]This is [**18**]; Borchardt was then the editor of the *Journal für die Reine und Angewandte Mathematik.*

[18]This is [**35**], but more specifically [**21**].

[19]This is [**22**], but we cite the English translation [**26**].

[20]This section introduces the key objects in play: the ring of integers $\mathcal{O}$ of a number field Ω, the order $\mathcal{O}' = \mathbb{Z}[\theta]$, and the index k.

[21]Dedekind says "finite field" for what we would call "a finite extension of $\mathbb{Q}$." He never considers fields with finitely many elements.

[22]Dedekind uses the lowercase fraktur $\mathfrak{o}$.

[23]Dedekind uses "ganzen Zahlen," literally "whole numbers," for algebraic integers. We will typically translate "integers." The elements of $\mathbb{Z}$ are "rational integers."

*I can only refer to the abstract. After several unsuccessful attempts to get it in the bookstore, I have recently obtained the original through the kindness of Professor Wangerin, but given my ignorance of the Russian language, to my great regret I was able to understand very little, only what is clear from looking at the formulas.

where

$$h_1, h_2, \dots, h_n$$

are arbitrary rational integers. The discriminant[24]

$$\Delta(\omega_1, \omega_2, \dots, \omega_n) = \Delta(\Omega) = D,$$

which is independent of the choice of the basis numbers[25] $\omega_1, \omega_2, \dots, \omega_n$, is called the fundamental number or the discriminant of the field Ω. (D. § 159, 160,162; B. § 12–18).

Now if θ is a specific algebraic integer in the field, we can set[26]

$$\begin{aligned}
1 &= c_1^{(0)}\omega_1 + c_2^{(0)}\omega_2 + \cdots + c_n^{(0)}\omega_n \\
\theta &= c_1^{(1)}\omega_1 + c_2^{(1)}\omega_2 + \cdots + c_n^{(1)}\omega_n \\
\theta^2 &= c_1^{(2)}\omega_1 + c_2^{(2)}\omega_2 + \cdots + c_n^{(2)}\omega_n \\
\dots &= \dots \\
\theta^{n-1} &= c_1^{(n-1)}\omega_1 + c_2^{(n-1)}\omega_2 + \cdots + c_n^{(n-1)}\omega_n
\end{aligned}$$

where all the n^2 coefficients or coordinates c are rational integers, and we will have

$$\Delta(1, \theta, \theta^2, \dots, \theta^{n-1}) = Dk^2,$$

where[27]

$$k = \sum \pm c_1^{(0)} c_2^{(1)} \dots c_n^{(n-1)}$$

is a rational integer. The absolute value of this number k, which is independent of the choice of integral basis, will for brevity from now one be called the *index* of the integer θ. If k is not 0, as we will always assume,[28] the numbers

$$1, \theta, \theta^2, \dots, \theta^{n-1}$$

will be independent of each other (D. § 159; B. § 4,15,17) and θ will be the root of an irreducible equation of degree n

$$F(\theta) = \theta^n + a_1\theta^{n-1} + \cdots + a_n = 0,$$

where the coefficients $1, a_1, a_2, \dots, a_n$ are all rational integers.

If we let $\varphi(t)$ be any function of the variable t, — and I remark that always, by this name [function] and by an expression of the form $\varphi(t)$, $f(t), \dots$ in this treatise one should always understand an entire function of t whose coefficients are rational integers[29] — the set $\mathscr{O}'$ of all other numbers of the form

$$\omega' = \varphi(\theta)$$

---

[24]The notion of the discriminant of a set of algebraic numbers seems to have been created by Dedekind by analogy to the older notion of the discriminant (or determinant) of a polynomial. In this paper Dedekind typically uses "Discriminante" for the general construct, reserving "Grundzahl," which we translate as "fundamental number," for this particular discriminant.

[25]Dedekind uses "Basiszahlen."

[26]Dedekind writes c_i' where we have $c_i^{(1)}$, c_i'' where we have $c_i^{(2)}$, etc.

[27]The meaning of this notation, standard at the time, is $k = \det[c_i^{(j)}]$. Dedekind does not use matrices, nor does he represent the determinant as an array.

[28]This running assumption is crucial, but it is not mentioned ever again. It is equivalent to assuming that θ is a generator.

[29]So "function" always means a polynomial with rational integer coefficients.

is called an *order* (D. § 165, 166; B. § 23); all such numbers are integers of the field Ω and therefore are contained also in $\mathcal{O}$. Clearly it suffices to take only the functions

$$\varphi(t) = x_0 + x_1 t + x_2 t^2 + \cdots + x_{n-1} t^{n-1}$$

whose degree is smaller than n, since if $\varphi_1(t)$ has degree larger than n we can divide it by

$$F(t) = t^n + a_1 t^{n-1} + a_2 t^{n-2} + \cdots + a_{n-1} t + a_n.$$

The remainder $\varphi(t)$ will have degree less than n and at the same time $\varphi_1(\theta) = \varphi(\theta)$. In the notation used above (B. § 3) we can set

$$\mathcal{O}' = [1, \theta, \theta^2, \ldots, \theta^{n-1}].$$

It also follows from the irreducibility of the equation $F(\theta) = 0$ that each number ω' can be represented in the form $\varphi(\theta)$[30] in only one way; nevertheless, in what follows we will not always restrict ourselves to that form of representation, but rather allow functions of any degree.

Prime numbers p — by which name we mean a rational positive prime number — fall in two cases once the fixed number θ is chosen:[31] the *first* case, which applies to infinitely many prime numbers, is when the index k of the number θ is *not* divisible by p. If $k = \pm 1$, then all primes are in this first case, and in fact $\mathcal{O}'$ is identical to $\mathcal{O}$. When however $k^2 > 1$, a finite number of primes will fall into the *second* case, namely the prime divisors of k. The paragraphs that follow will show that the decomposition of the prime numbers p of the first kind (or rather the decomposition of the corresponding principal ideals* $\mathcal{O}p$) as a product of prime ideals[32] can be completely reduced to the decomposition of the function $F(t)$ as a product of functions that are prime[33] with respect to the modulus p (C. 6).[34] On the other hand, it is not possible to do this in the same simple way for prime numbers of the second kind. The following remarks should be made before this investigation.

Let p be a fixed prime of the *first* kind, so that k is not divisible by p. In this case an element of $\mathcal{O}'$

$$\omega' = x_0 + x_1\theta + x_2\theta^2 + \cdots + x_{n-1}\theta^{n-1}$$

is divisible by p (which means it is equal to $p\omega$, with ω an integer, i.e., an element of $\mathcal{O}$) if and only if all the coefficients x_i are divisible by p. To see that, notice[35] that

$$\omega' = h_1\omega_1 + h_2\omega_2 + \cdots + h_n\omega_n$$

---

[30]Dedekind means $\varphi(t)$ with φ of degree less than n.

[31]The idea, then, is to choose and fix $\theta \in \mathcal{O}$ such that $\Omega = \mathbb{Q}(\theta)$. Then $\mathcal{O}' = \mathbb{Z}[\theta] \subset \mathcal{O}$ and $k = (\mathcal{O} : \mathcal{O}')$ is the index. The rational primes p that divide k are those in the second case; the (infinitely many) other primes are in the first case.

[32]Here and elsewhere Dedekind writes "Produkte aus lauter Primidealen," literally "product of nothing but prime ideals," to emphasize that it is a complete factorization into primes.

[33]We would say irreducible modulo p.

[34]This is the theorem announced in the *Anzeige* **[20]**, translated above.

[35]This paragraph is simple linear algebra. Dedekind uses the expression of the powers of θ in terms of the integral basis to rewrite ω' in terms of the basis.

*This notation for principal ideals is more appropriate than $\mathfrak{i}(p)$, which I used earlier (D. § 163).

where

$$\begin{aligned} h_1 &= c_1^{(0)}x_0 + c_1^{(1)}c_1 + c_1^{(2)}x_2 + \cdots + c_1^{(n-1)}x_{n-1} \\ h_2 &= c_2^{(0)}x_0 + c_2^{(1)}c_1 + c_2^{(2)}x_2 + \cdots + c_2^{(n-1)}x_{n-1} \\ &\cdots\cdots \\ h_n &= c_n^{(0)}x_0 + c_n^{(1)}c_1 + c_n^{(2)}x_2 + \cdots + c_n^{(n-1)}x_{n-1} \end{aligned}$$

It follows from the independence of $\omega_1, \omega_2, \ldots, \omega_n$ that ω' is divisible by p if and only if each of the coordinates $h_1, h_2, \ldots, h_n$ is divisible by p. If so, each of the products $kx_0, kx_1, kx_2, \ldots, kx_{n-1}$ is also divisible by p, and therefore so are the coefficients $x_0, x_1, x_2, \ldots, x_{n-1}$.[36] The same theorem can clearly also be stated as: a number ω' of the order $\mathcal{O}'$ is divisible by a prime number p of the first kind if the quotient $\frac{\omega'}{p}$ is itself in the order $\mathcal{O}'$. Conversely, when all the coefficients $x_0, x_1, x_2, \ldots, x_{n-1}$ are all divisible by p, then obviously ω' is divisible by p.[37]

Therefore[38] two numbers $\varphi_1(\theta)$ and $\varphi_2(\theta)$ of the order $\mathcal{O}'$ are congruent modulo p (i.e., their difference $\varphi_1(\theta) - \varphi_2(\theta)$ is divisible by p) if and only if the coefficients of the two functions $\varphi_1(t)$ and $\varphi_2(t)$ are all congruent modulo p, i.e., in the sense of the theory of higher congruences, when we have

$$\varphi_1(t) \equiv \varphi_2(t) \pmod p$$

(C. 1). For this conclusion, however, we need to assume that the degrees of the functions $\varphi_1(t)$ and $\varphi_2(t)$ are less than n. If that is not the case, after dividing by $F(t)$ we obtain an identity of the form

$$\varphi_1(t) - \varphi_2(t) = F(t)\psi(t) + \psi_1(t),$$

where $\psi_1(t)$ has degree less than n, and then $\varphi_1(\theta) - \varphi_2(\theta) = \psi_1(\theta)$. We will have

$$\varphi_1(\theta) \equiv \varphi_2(\theta) \pmod p$$

when $\psi_1(t) = p\psi_2(t)$, that is, when

$$\varphi_1(t) - \varphi_2(t) = F(t)\psi(t) + p\psi_2(t).$$

The existence of such an identity is described in the theory of higher congruences as

$$\varphi_1(t) - \varphi_2(t) \equiv F(t)\psi(t) \pmod p$$

or simply as (C. 7) as

$$\varphi_1(t) \equiv \varphi_2(t) \pmod{p, F(t)}.$$

Conversely, it is clear that from that function congruence the number congruence

$$\varphi_1(\theta) \equiv \varphi_2(\theta)$$

also follows; the two congruences are therefore equivalent. Thus in $\mathcal{O}'$ there are as many numbers $\varphi(\theta)$ that are incongruent modulo p as there are functions $\varphi(t)$ incongruent with respect to the double modulus $p, F(t)$; there are p^n of the latter (C. 8), which is also the number[39] $(\mathcal{O} : \mathcal{O}p) = N(p)$ of numbers in $\mathcal{O}$ that are

[36] If the matrix $[c_i^j]$ is invertible mod p, then $[c_i^j]\vec{x} = \vec{h} \equiv \vec{0}$ if and only if $\vec{x} \equiv \vec{0}$.

[37] Added a paragraph break here.

[38] Dedekind will now translate congruences between elements $\varphi(\theta)$ of $\mathbb{Z}[\theta]$ into congruences between the polynomials $\varphi(t)$. He does so first under the assumption that $\deg(\varphi) < n$, and then in the general case. For the latter he uses the "double modulus" $p, F(t)$.

[39] Dedekind writes $(\mathcal{O}, \mathcal{O}p)$ for the index.

incongruent modulo p (B. § 18; D. § 162), which implies the following result: *each number ω of the domain $\mathcal{O}$ is congruent modulo p to a number ω' of the order $\mathcal{O}'$.*[40]

The same conclusion can be reached directly by the following simple argument. From the n relations between the numbers $1, \theta, \theta^2, \ldots, \theta^{n-1}$, on the one hand, and the numbers $\omega_1, \omega_2, \ldots, \omega_n$, on the other, it follows that the products $k\omega_1, k\omega_2, \ldots, k\omega_n$ are contained in the order $\mathcal{O}'$, and therefore so are all the products $k\omega$ for any ω belonging to $\mathcal{O}$. Therefore we have $k\omega = \varphi(\theta)$. Now since k is not divisible by p, we can choose a rational integer l such that $kl \equiv 1 \pmod{p}$, and then it follows that $\omega \equiv lk\omega \equiv l\varphi(\theta) \pmod{p}$, so that ω is congruent modulo p to a number $l\varphi(\theta)$ that is in the order $\mathcal{O}'$.

Things are completely different when p is a prime the *second* kind.[41] In that case the determinant k is divisible by p, and it is easy to prove that there exist n rational integers $x_0, x_1, \ldots, x_{n-1}$, not all divisible by p, such that the corresponding numbers $h_1, h_2, \ldots, h_n$ are all divisible by p. Then the corresponding number

$$\omega' = x_0 + x_1\theta + x_2\theta^2 + \cdots + x_{n-1}\theta^{n-1}$$

is in fact divisible by p even though the coefficients $x_0, x_1, \ldots, x_{n-1}$ are not all divisible by p. It follows that the number $(\mathcal{O}' : \mathcal{O}p)$ of incongruent elements in $\mathcal{O}'$ is smaller than p^n.[42] It follows that there are numbers ω in $\mathcal{O}$ that are not congruent modulo p to *any* element $\varphi(\theta)$, i.e., there exist congruence classes (mod p) in $\mathcal{O}$ for which there is no representative in $\mathcal{O}'$. The precise determination of the number $(\mathcal{O}' : \mathcal{O}p)$ is not necessary for our purposes.[43]

§2

In this paragraph[44] we consistently make the assumption that p is a prime number of the *first* kind. We want to prove that in this case the theory of higher congruences gives an easy way to decompose a principal ideal $\mathcal{O}p$ into its prime factors. This happens because the function $F(t)$, which we will denote F for brevity, factors modulo p as a product of *prime functions* $P(t)$ (C. 6). If we assume, for convenience, that each prime function P has highest coefficient $= 1$, it follows that two incongruent prime functions are always relatively prime (C. 5). Combining all the congruent factors into powers we get

$$F \equiv P_1^{e_1} P_2^{e_2} \ldots P_m^{e_m} \pmod{p}$$

[40]What Dedekind has shown is that when p does not divide the index the quotient $\mathcal{O}/p\mathcal{O}$ is isomorphic to $\mathcal{O}'/p\mathcal{O}'$. As he shows next, if k is the index $(\mathcal{O} : \mathcal{O}')$ the isomorphism is given by multiplication by lk, where l is any rational integer such that $lk \equiv 1 \pmod{p}$. This is a crucial move, since Dedekind has no direct way to work with congruences between elements of $\mathcal{O}$. By contrast, congruences between elements of $\mathcal{O}'$ translate directly into polynomial congruences, which will allow Dedekind to use the results of **[18]**.

[41]For index divisors, $\mathcal{O}'/p\mathcal{O}' \hookrightarrow \mathcal{O}/p\mathcal{O}$ is not onto.

[42]We would write $(\mathcal{O}' : \mathcal{O}'p)$, but Dedekind does not.

[43]The editors of **[25]** add a footnote here: "In Zolotareff one also finds the theorem that the exceptional prime numbers are precisely those for which there is a number ω' in the order $\mathcal{O}'$ that are divisible by p but whose coefficients are not all divisible by p. Zolotareff does not say, however, that these prime numbers are the index divisors."

[44]This section states and proves "Dedekind's Theorem," describing the factorization of primes of the first kind in terms of higher congruences. Dedekind will use the phrase "prime function" to mean a monic polynomial that is irreducible modulo p. Everything in this section assumes that p does not divide k.

where the P_i are all the incongruent prime functions contained in F.[45]

Let P be any one of these m prime functions, and let $\rho = P(\theta)$. Then there is an ideal $\mathfrak{p}$ that is the greatest common divisor of $\mathscr{O}p$ and $\mathscr{O}\rho$. To study the properties of this ideal $\mathfrak{p}$, we first determine all the elements $\varphi(\theta)$ contained in the order $\mathscr{O}'$ that are divisible by $\mathfrak{p}$ (i.e., are contained in $\mathfrak{p}$). We want to prove[46] that the congruence

$$\psi(\theta) \equiv 0 \pmod{\mathfrak{p}} \tag{1}$$

is completely equivalent to the function congruence

$$\psi(t) \equiv 0 \pmod{p, P}. \tag{2}$$

Indeed,[47] by definition (D. § 163; B. § 19) the ideal $\mathfrak{p}$ is the collection of all numbers of the form

$$\rho\alpha + p\beta,$$

where α, β are arbitrary numbers from the domain $\mathscr{O}$. By §1, each number α is congruent modulo p to some number $\varphi(\theta)$ in the order $\mathscr{O}'$, so from (1) we get a congruence of the form

$$\psi(\theta) \equiv P(\theta)\varphi(\theta) \pmod{p};$$

this is equivalent (as in §1) to the function congruence

$$\psi(t) \equiv P(t)\varphi(t) \pmod{p, F},$$

and therefore also equivalent to congruence (2), since F is divisible by P. Conversely, it follows immediately[48] from (2) that any $\psi(\theta)$ is of the form $\rho\alpha + p\beta$, and so is $\equiv 0 \pmod{\mathfrak{p}}$ as well. This proves our claim above.[49]

With the help of these results we can easily[50] compute the *norm* of the ideal $\mathfrak{p}$, i.e., the number $(\mathscr{O} : \mathfrak{p}) = N(\mathfrak{p})$ of elements of $\mathscr{O}$ that are incongruent modulo $\mathfrak{p}$. So let α_1, α_2 be any two numbers in $\mathscr{O}$. From §1 we now that there exist two numbers $\varphi_1(\theta)$, $\varphi_2(\theta)$ in $\mathscr{O}'$ that are congruent modulo p to α_1, α_2. Since $\mathfrak{p}$ divides p, we also have

$$\alpha_1 \equiv \varphi_1(\theta), \quad \alpha_2 \equiv \varphi_2(\theta) \pmod{\mathfrak{p}}.$$

So the two numbers α_1, α_2 are congruent modulo $\mathfrak{p}$ if and only if

$$\varphi_1(\theta) \equiv \varphi_2(\theta) \pmod{\mathfrak{p}}.$$

This congruence is equivalent, as above, to the congruence

$$\varphi_1(t) \equiv \varphi_2(t) \pmod{p, P}.$$

[45] Dedekind knows that there is unique factorization in $\mathbb{F}_p[x]$. This is one of the many results in **[18]**.

[46] Dedekind doesn't state this as a separate Lemma but he uses it over and over in the sequel.

[47] Here begins the proof. Recall that $\rho = P(\theta)$ where P is an irreducible factor of F.

[48] From (2) we get $\psi(t) = P(t)\varphi(t) + pG(t)$; plugging in θ gives $\psi(\theta) = \rho\varphi(\theta) + pG(\theta) \in \mathfrak{p}$, since $\varphi(\theta) \equiv \alpha \pmod{p}$.

[49] The lemma is now proved.

[50] In modern terms, the argument is to pass from $\mathscr{O}/\mathfrak{p}$ to $\mathscr{O}'/\mathfrak{p}$ and then to translate congruences between elements of $\mathscr{O}'$ into congruences of polynomials using the Lemma above. This reduces the problem to counting incongruent polynomials modulo p, P, which Dedekind had already done in **[18]**.

Therefore there are as many numbers α that are incongruent modulo $\mathfrak{p}$ as there are functions $\varphi(t)$ incongruent with respect to the double modulus p, P; this quantity is $= p^f$, where f is the degree of the function P (C. 8), so we have

$$N(\mathfrak{p}) = p^f.$$

With that it is easy to prove $\mathfrak{p}$ is a *prime ideal.* First, we know $f \geq 1$, so $N(\mathfrak{p}) \neq 1$, so $\mathfrak{p}$ cannot be equal to $\mathcal{O}$. It suffices then to show[51] that $\mathfrak{p}$ is not a decomposable ideal, i.e., that it is not a product of the form $\mathfrak{a}_1\mathfrak{a}_2$, where $\mathfrak{a}_1\mathfrak{a}_2$ are ideals and neither is equal to $\mathcal{O}$. Such a decomposable[52] ideal $\mathfrak{m} = \mathfrak{a}_1\mathfrak{a}_2$ has the characteristic property that there are always two numbers α_1, α_2, neither divisible by $\mathfrak{m}$, whose product $\alpha_1\alpha_2$ is divisible by $\mathfrak{m}$. This is because both the ideals $\mathfrak{a}_1, \mathfrak{a}_2$ are different from $\mathcal{O}$, so neither of them can be divisible by their product $\mathfrak{m} = \mathfrak{a}_1\mathfrak{a}_2$. So there must exist a number α_1 that is divisible by $\mathfrak{a}_1$ but not by $\mathfrak{m}$, and similarly an α_2 that is divisible by $\mathfrak{a}_2$ but not by $\mathfrak{m}$. So $\mathfrak{p}$ will be a prime ideal if we can show[53] that a product $\alpha_1\alpha_2$ cannot be divisible by $\mathfrak{p}$ unless at least one of the factors α_1, α_2 is divisible by $\mathfrak{p}$. For this,[54] we set, as above,

$$\alpha_1 \equiv \varphi_1(\theta), \alpha_2 \equiv \varphi_2(\theta) \pmod{\mathfrak{p}},$$

so that

$$\alpha_1\alpha_2 \equiv \varphi_1(\theta)\varphi_2(\theta) \pmod{\mathfrak{p}},$$

and since $\alpha_1\alpha_2 \equiv 0 \pmod{\mathfrak{p}}$, we must have

$$\varphi_1(\theta)\varphi_2(\theta) \equiv 0 \pmod{\mathfrak{p}}$$

and so

$$\varphi_1(t)\varphi_2(t) \equiv 0 \pmod{p, P}.$$

Since P is a *prime function* it follows[55] that one of the two congruences

$$\varphi_1(t) \equiv 0 \pmod{p, P} \qquad \text{or} \qquad \varphi_2(t) \equiv 0 \pmod{p, P}$$

must hold (C. 6). So at least one of the congruences

$$\varphi_1(\theta) \equiv 0 \pmod{\mathfrak{p}} \qquad \text{or} \qquad \varphi_2(\theta) \equiv 0 \pmod{\mathfrak{p}}$$

must be true, that is, one of the two numbers α_1, α_2 must $\equiv 0 \pmod{\mathfrak{p}}$. Therefore $\mathfrak{p}$ is a prime ideal, and we know (B. § 21) that $\mathfrak{p}$ is a prime ideal of *degree* f, since $N(\mathfrak{p}) = p^f$.

Now[56] we would like to prove that the highest exponent e of P in the factorization of F and the highest exponent of $\mathfrak{p}$ in the factorization of p are equal. Indeed, if F is divisible modulo p by P^e but not by P^{e+1}, We have

$$F \equiv SP^e \pmod{p},$$

[51]Here we see that (at this time) Dedekind's working definition of "prime ideal" is not the same as the one we learn today. The next several lines explain why it is enough to prove that $\alpha_1\alpha_2 \in \mathfrak{p}$ implies that either α_1 or α_2 is in $\mathfrak{p}$. The argument is straightforward; note that Dedekind consistently writes "$\mathfrak{m}$ divides α" instead of "α belongs to $\mathfrak{m}$."

[52]The proof starts here.

[53]We have shown that an ideal I is indecomposable if and only if $ab \in I$ implies either $a \in I$ or $b \in I$.

[54]Now we will prove $\mathfrak{p}$ is prime; as usual, we reduce to elements of $\mathcal{O}'$ and then to polynomial congruences.

[55]Irreducibles in $\mathbb{F}_p[t]$ are prime, which Dedekind had proved in **[18]**.

[56]We have a prime ideal dividing p, so it remains to determine the valuation, i.e., the highest power of $\mathfrak{p}$ dividing p.

where S is not divisble by P. It follows as above that

$$\sigma = S(\theta)$$

is not divisible by $\mathfrak{p}$. Since $\mathfrak{p}$ is the greatest common divisor of $\mathscr{O}p$ and $\mathscr{O}\rho$, we know that

$$\mathscr{O}p = \mathfrak{p}\mathfrak{a}, \qquad \mathscr{O}\rho = \mathfrak{p}\mathfrak{b}$$

with $\mathfrak{a}$ and $\mathfrak{b}$ relatively prime. So what we need to prove[57] is that the highest power of $\mathfrak{p}$ contained in $\mathfrak{a}$ is $\mathfrak{p}^{e-1}$. For this, consider the number

$$\eta = \sigma\rho^{e-1} = S(\theta)P(\theta)^{e-1},$$

which cannot be divisible by p, since the degree of the polynomial SP^{e-1} is less than n and its highest coefficient is $= 1$. On the other hand, η is divisible by $\mathfrak{p}^{e-1}$, since ρ is divisible by $\mathfrak{p}$. From the congruence $F \equiv SP^e \pmod{p}$, we see that $\eta\rho = \sigma\rho^e$ is divisible by p. So the ideal $\eta\mathfrak{p}\mathfrak{b}$ is divisible by $\mathfrak{p}\mathfrak{a}$, and therefore $\eta\mathfrak{b}$ is divisible by $\mathfrak{a}$; since $\mathfrak{a}$ and $\mathfrak{b}$ are relatively prime, we see that η is divisible by $\mathfrak{a}$. So let

$$\mathscr{O}\eta = \mathfrak{a}\mathfrak{c},$$

where $\mathfrak{c}$ is an ideal not divisible by $\mathfrak{p}$,* because otherwise η would be divisible by $\mathfrak{a}\mathfrak{p} = \mathscr{O}p$, which we know is not the case. Since η is divisible by $\mathfrak{p}^{e-1}$, so is $\mathfrak{a}$.[58]

We now only need to show that $\mathfrak{a}$ is not divisible by $\mathfrak{p}^e$. Since[59] $e \geq 1$, if $\mathfrak{a}$ is divisible by $\mathfrak{p}^e$, then it is certainly divisible by $\mathfrak{p}$ itself. Now if $\mathfrak{a}$ is divisible by $\mathfrak{p}$, $\mathfrak{b}$ cannot be divisible by $\mathfrak{p}$, and therefore ρ is not divisible by $\mathfrak{p}^2$. From that it follows that σ is not divisible by $\mathfrak{p}$, so in this case $\mathfrak{p}^{e-1}$ is the highest power of $\mathfrak{p}$ contained in the number $\eta = \sigma\rho^{e-1}$. So η, and therefore the ideal $\mathfrak{a}$ contained in it, cannot be divisible by $\mathfrak{p}^e$, which was to be proved.

After this the investigation of a specific[60] prime function P contained in F and its corresponding prime ideal $\mathfrak{p}$ is complete. We now apply the results to all the functions contained in F,

$$F \equiv P_1^{e_1} P_2^{e_2} \dots P_m^{e_m} \pmod{p},$$

with incongruent prime functions

$$P_1, P_2, \dots, P_m$$

---

[57] Dedekind first proves that $\mathfrak{a}$ is divisible by $\mathfrak{p}^{e-1}$, and then proves it cannot be divisible by $\mathfrak{p}^e$.

[58] Added a paragraph break here. The first part of the proof is finished: $\mathfrak{a}$ is divisible by $\mathfrak{p}^{e-1}$ and so $\mathscr{O}p = \mathfrak{p}\mathfrak{a}$ is divisible by $\mathfrak{p}^e$; in valuation terms, $v_{\mathfrak{p}}(\mathfrak{a}) \geq e - 1$.

[59] The argument opens with "if $\mathfrak{a}$ is divisible by $\mathfrak{p}^e$," which seems like setting up a proof by contradiction. But that is not where the argument goes. We think it is better expressed as two cases: if $\mathfrak{a}$ is not divisible by $\mathfrak{p}$, then since $0 = v_{\mathfrak{p}}(\mathfrak{a}) \geq e - 1$ we must have $e = 1$ and we are done. If $\mathfrak{a}$ is divisible by $\mathfrak{p}$, then Dedekind shows that $v_{\mathfrak{p}}(\rho) = 1$ and so the number $\eta = \sigma\rho^{e-1}$ is divisible by $\mathfrak{a}$ but not by $\mathfrak{p}^e$. From that it follows that $v_{\mathfrak{p}}(\mathfrak{a}) \leq e - 1$, hence that $v_{\mathfrak{p}}(\mathfrak{a})$ must be exactly $e - 1$.

[60] For each different irreducible factor P we have found a prime ideal $\mathfrak{p}$ dividing p and shown that the multiplicity of P as a factor of F is the same as the multiplicity of $\mathfrak{p}$ as a factor of p. To complete the proof we put these all together. Two crucial things need to be proved: that the prime ideals corresponding to two different irreducible factors are not the same (so that p is divisible by their product) and that these are no other prime ideals dividing p.

*It follows that $\mathfrak{a}$ is the greatest common divisor of the ideals $\mathscr{O}p$ and $\mathscr{O}\eta$, and so $\eta\mathfrak{p}$ is the least common multiple of $\mathscr{O}p$ and $\mathscr{O}\eta$, i.e., $\mathfrak{p}$ is the collection of all roots π of the congruence $\eta\pi \equiv 0 \pmod{p}$. This could also have been used to define the ideal $\mathfrak{p}$.

of degrees, respectively,

$$f_1, f_2, \ldots, f_m.$$

To these functions correspond prime ideals

$$\mathfrak{p}_1, \mathfrak{p}_2, \ldots, \mathfrak{p}_m$$

with the corresponding degrees, so that

$$N(\mathfrak{p}_1) = p^{f_1}, N(\mathfrak{p}_2) = p^{f_2}, \ldots, N(\mathfrak{p}_m) = p^{f_m}$$

and

$$\mathfrak{p}_1^{e_1}, \mathfrak{p}_2^{e_2}, \ldots, \mathfrak{p}_m^{e_m}$$

are the highest powers of these ideals contained p. These m ideals are all distinct;[61] for example, P_2 is not divisible by P_1 mod p, so the number $P_2(\theta)$ is divisible by $\mathfrak{p}_2$ but not by $\mathfrak{p}_1$, and it follows that $\mathfrak{p}_1$ and $\mathfrak{p}_2$ are different ideals. Finally,[62] we know that p cannot be divisible by any other prime ideal, since[63]

$$P_1(\theta)^{e_1} P_2(\theta)^{e_2} \ldots P_m(\theta)^{e_m} \equiv 0 \pmod{p}.$$

If p is divisible by a prime ideal, that ideal has to divide one of the m numbers $\rho = P(\theta)$; but then that ideal must be identical to the prime ideal $\mathfrak{p}$, which is the greatest common divisor of $\mathcal{O}p$ and $\mathcal{O}\rho$.

From all this it follows (D. § 163, B. § 25) that

$$\mathcal{O}p = \mathfrak{p}_1^{e_1} \mathfrak{p}_2^{e_2} \ldots \mathfrak{p}_m^{e_m}.$$

A consequence[64] of this, found by taking norms, is

$$n = e_1 f_1 + e_2 f_2 + \cdots + e_m f_m.$$

Thus we have proved the following theorem,[65] which I announced in the *Göttingischen gelehrten Anziegen* in September 20, 1871.[66]

[I] *Let k be the index*[67] *of the number θ that satisfies the irreducible equation $F(\theta) = 0$ of degree N. If k is not divisible by p and if*

$$F \equiv P_1^{e_1} P_2^{e_2} \ldots P_m^{e_m} \pmod{p}$$

where the $P_1, P_2, \ldots, P_m$ are incongruent prime functions of degree $f_1, f_2, \ldots, f_m$, respectively, then we have

$$\mathcal{O}p = \mathfrak{p}_1^{e_1} \mathfrak{p}_2^{e_2} \ldots \mathfrak{p}_m^{e_m},$$

where $\mathfrak{p}_1, \mathfrak{p}_2, \ldots, \mathfrak{p}_m$ are pairwise distinct prime ideals whose degrees are, respectively, $f_1, f_2, \ldots, f_m$, and for each distinct prime function P the corresponding prime ideal $\mathfrak{p}$ is the greatest common divisor of the ideals $\mathcal{O}p$ and $\mathcal{O}P(\theta)$.

61This is a key fact later on: distinct factors correspond to distinct ideals and vice versa. Dedekind shows only that different P give different ideals.

62The last thing to note is that we have the complete factorization: no other prime ideals divide p. This is easy to see.

63The product of all the $P_i(\theta)^{e_i}$ is congruent mod p to $F(\theta) = 0$.

64Surprisingly, this famous formula does not appear in **[21]** or **[26]**.

65Usually known today as "Dedekind's theorem" or "Dedekind's factorization theorem."

66Dedekind does not use the label "Theorem." The main theorems are marked by roman numerals.

67Recall the running assumption that $k \neq 0$, so that $\Omega = \mathbb{Q}(\theta)$ and the miminal polynomial is of degree n.

§3

From this theorem it follows that on the basis of a specific integer θ from the field Ω, which allows one to represent as $\varphi(\theta)$ infinitely many integers,[68] one can find the factorization of all the prime numbers p that do not not divide the index corresponding to a the chosen θ. It is therefore very important to know whether a prime number p is a divisor of the index k or not.[69] If we have a basis $\omega_1, \omega_2, \ldots, \omega_n$ of the domain $\mathscr{O}$, or even just know the fundamental number D of the field Ω, it is easy to answer the question, since in that case we can find k directly. From the coefficients of the equation $F(\theta) = 0$ we can compute its discriminant

$$\Delta(1, \theta, \theta^2, \ldots, \theta^{n-1}) = (-1)^{\frac{1}{2}n(n-1)} N(F'(\theta)) = Dk^2,$$

and from that we can find the square of the index k by dividing by D. In most investigations, however, things are very different, since only the equation $F(\theta) = 0$ is known, and not the fundamental number D of the corresponding field Ω. We would like to decide on that basis[70] whether or not a specific prime number p divides the unknown index of the number θ. This is in fact possible, as we will now show, with the help of the theory of higher congruences. Using our previous notation, the answer turns out to depend on the nature of the function[71] M that appears in the identity

$$F = P_1^{e_1} P_2^{e_2} \ldots P_m^{e_m} - pM.$$

This will be the content of the next two theorems.[72]

---

[68]While it is certainly true that $\mathbb{Z}[\theta]$ is an infinite set of algebraic integers, that is true for any element $\theta \in \mathscr{O}$, so it cannot be what Dedekind actually means here. Perhaps he just means to restate his running assumption that $k \neq 0$, which implies that $\mathbb{Z}[\theta]$ has finite index in $\mathscr{O}$.

[69]The goal of this section is to characterize the primes p that divide the index of θ. Dedekind points out that this is easy if we have the discriminant D but his goal is to answer the question solely in terms of the minimal polynomial F.

[70]That is, solely on the basis of the equation.

[71]This equation defines the polynomial M: it is the result of dividing the difference by p. Since F is congruent mod p to the product, M is a polynomial with rational integer coefficients.

[72]The first theorem specifies a property of M modulo p, but choosing different lifts for the factors P_i can change M (even mod p). So we need to check that the divisibility property we are looking for is independent of the chosen lifts.

Lemma: Suppose $P, R, S, T \in \mathbb{Z}[t]$, $P \equiv R \pmod{p}$, $S \equiv T \pmod{p}$, P is irreducible mod p, and that

$$F = P^e S - pM = R^e T - pN$$

with $e \geq 2$. Then $M - N$ is divisible by P mod p.

Proof of Lemma: The equation $P^e S - pM = R^e T - pN$ gives

$$M - N = \frac{1}{p}(P^e S - R^e T),$$

so what we need to show is $P^e S \equiv R^e T \pmod{p^2}$. Writing $R = P + pX$, $T = S + pY$ we get

$$P^e S - R^e T \equiv P^e S - (P^e + epP^{e-1}X)(S + pY) \equiv p(eP^{e-1}S + P^e Y) \pmod{p^2}.$$

Dividing by p gives

$$M - N \equiv eP^{e-1}S + P^e Y \pmod{p},$$

and since $e \geq 2$ we are done. □

Dedekind does not prove this lemma; rather, he *deduces* it from the fact that the question of whether the index is divisible by p is independent of the choice of lifts. But he does say it can be checked directly, and it seems better to do that.

[II] *If the index of the number θ is not divisible by p, then M cannot be divisible mod p by any prime function P whose square divides F mod p.*

To prove this, we can use the results in the previous paragraph, which were all obtained under the assumption that p does not divide k. Retaining the same notation we used there,[73] write $F \equiv SP^e \pmod{p}$, or

$$F = SP^e - pM,$$

and suppose $e \geq 2$. Then p is divisible by $\mathfrak{p}^2$, so $\mathfrak{a}$ is divisible by $\mathfrak{p}$ and $\mathfrak{b}$ is not.[74] Therefore,[75] $\mathfrak{p}^e$ is the highest power of $\mathfrak{p}$ dividing $S(\theta)P(\theta)^e = pM(\theta)$. Since p is divisible[76] by $\mathfrak{p}^e$, it follows that $M(\theta)$ cannot be divisible by $\mathfrak{p}$ and so $M \not\equiv 0 \pmod{p, P}$, as claimed.[77]

It is also possible to prove the theorem without using the results in the previous paragraph, in following indirect but equivalent form:

If F is divisible mod p by the square of an irreducible polynomial P, so that $F = SP^e - pM$ with $e \geq 2$, and M is divisible by P, then the index k of the number θ will be divisible by p.

Let the letters ρ, σ, η have the same meanings as in the previous paragraph, so that we set

$$\rho = P(\theta), \qquad \sigma = S(\theta), \qquad \eta = \sigma\rho^{e-1}.$$

Using the results of §1,[78] the proof of our theorem will be complete if we can show that the number $\eta = S(\theta)P(\theta)^{e-1}$ must be divisible by p, since the function SP^{e-1} is of degree lower than n and not $\equiv 0 \pmod{p}$. To prove that η is divisible by p, it suffices to show that each power of a prime ideal dividing p also divides η (D. § 163, B. § 25). To this end set

$$\mu = M(\theta);$$

consider the equation

$$\sigma\rho^e = \eta\rho = p\mu.$$

First, if $\mathfrak{p}$ is a prime ideal dividing p but not dividing ρ, then from $\eta\rho = p\mu$ it follows at once that η is divisible by the highest power of $\mathfrak{p}$ dividing p. Next, suppose $\mathfrak{p}$ divides both p and ρ. Since S and P are relatively prime functions,[79] there exist (C. 4) two functions U, V such that the congruence

$$SU + PV \equiv 1 \pmod{p}$$

holds. From that we get the numerical congruences[80]

$$\sigma U(\theta) + \rho V(\theta) \equiv 1 \pmod{p}$$

[73]So S is the product of all the irreducible factors different from P; in particular, S is not divisible by P.

[74]Since $\mathfrak{a}$ and $\mathfrak{b}$ are relatively prime.

[75]Since the principal ideal $\mathcal{O}\rho$ is equal to $\mathfrak{p}\mathfrak{b}$, we know that $\rho = P(\theta)$ is divisible by $\mathfrak{p}$ only once, and that $\sigma = S(\theta)$ is not divisible by $\mathfrak{p}$.

[76]Dedekind means $\mathfrak{p}^e$ is the highest power of $\mathfrak{p}$ that divides p, of course.

[77]This concludes the proof of Theorem II.

[78]If p does not divide the index k, then two elements of $\mathcal{O}'$ are congruent mod p if and only if the corresponding polynomials are congruent mod p, F. In our case the polynomial will have degree less than n, so being congruent mod p, F is equivalent to being congruent mod p.

[79]Dedekind doesn't say so, but he means relatively prime mod p.

[80]For the first one, we just plug in θ; for the second, remember that $\mathfrak{p}$ divides both p and ρ.

$$\sigma U(\theta) \equiv 1 \pmod{\mathfrak{p}}.$$

and it follows that σ is not divisible by $\mathfrak{p}$. Let[81] $\mathfrak{p}^h$, $\mathfrak{p}^r$, $\mathfrak{p}^m$ be the highest powers of $\mathfrak{p}$ dividing p, ρ, μ, respectively. Since $\sigma\rho^e = p\mu$ and $\eta = \sigma\rho^{e-1}$, we see that

$$er = h + m,$$

and also that the highest power of $\mathfrak{p}$ appearing in η is equal to

$$(e-1)r = h + m - r.$$

Since we want to show that η is divisible by $\mathfrak{p}^h$, it remains to prove that

$$m \geq r.$$

Now we have to consider two cases.[82] In the first, $r \geq h$, we use the first assumption of our theorem, namely that $e \geq 2$. Then $h+m = er \geq 2r$, and so $m-r \geq r-h \geq 0$, as claimed. In the second case, $r \leq h$, we use the second assumption in our theorem, namely that $M \equiv 0 \pmod{p, P}$, i.e., $M \equiv PT \pmod{p}$. Therefore $\mu \equiv \rho T(\theta) \pmod{p}$. Since ρ is divisible by $\mathfrak{p}^r$, it follows from this congruence that μ is also divisible by $\mathfrak{p}^r$, and so that $m \geq r$, as we wanted to prove.

Now that we have proved Theorem II in two different ways, we will also show the correctness of the converse.

[III] *If M is not divisible mod p by any prime function P whose square divides F mod p, the index k of the number θ is not divisible by p.*

The same theorem clearly can also be stated in the following form:

If the index k of a number θ is divisible by p, there exists a prime function P dividing M whose square divides F modulo p.

We present the proof[83] of the latter form [of the theorem], because the assumption that k is divisible by p is easier to use, insofar as (according to §1) it implies the existence of a number

$$\varphi(\theta) = x_0 + x_1\theta + x_2\theta^2 + \cdots + x_{n-1}\theta^{n-1}$$

which is divisible by p but whose coefficients $x_0, x_1, x_2, \ldots, x_{n-1}$ are not all[84] divisible by p. Let us first denote by A the greatest common divisor of $\varphi(t)$ and F modulo p. The degree of A is smaller than n, since φ has degree smaller than n, and it is also not $\equiv 0 \pmod{p}$. Write

$$F = AB - pM,$$

[81] From here on we are basically computing $\mathfrak{p}$-adic valuations.

[82] The two cases are $r \geq h$ and $r \leq h$. Dedekind will use the assumption that $e \geq 2$ to handle the first case and the assumption that M is divisible by P to handle the second.

[83] The structure of the proof is as follows. If $p|k$ then there exists a polynomial $\varphi(t) \in \mathbb{Z}[t]$ such that $\varphi(t) \not\equiv 0 \pmod{p}$ but $\varphi(\theta)$ is divisible by p in $\mathcal{O}$. We look at $A = \gcd(F, \varphi)$ (over $\mathbb{F}_p$, but choose a monic lift of degree $< n$) and set $F = AB - pM$. Then we show that any prime divisor P of B in $\mathbb{F}_p[t]$ also divides M, therefore divides F, and we can then show it divides A as well, so that $P^2|F$. Factoring out the largest power of P gives $F = P^e A'B' - pM$ with $P \nmid A'B'$, $e \geq 2$, $P|M$, which is what we want. Notice that the polynomial denoted by M might change in the course of the argument.

[84] We can say something a little stronger that will help below: if we had all but x_0 divisible by p, then $\varphi(\theta) \equiv x_0 \pmod{p}$ and so x_0 is also divisible by p. This means that $\varphi(t)$ is not a constant mod p.

so that B is not a constant.[85] There exist (C. 4) two functions φ_1, φ_2 such that

$$\varphi(t)\varphi_1(t) + F(t)\varphi_2(t) \equiv A(t) \pmod{p}.$$

From this it follows[86] that the number $A(\theta)$ is also divisible by p,* From that[87] we get an equation of the form

$$A(\theta)^s + ph_1 A(\theta)^{s-1} + \cdots + p^s h^s = 0,$$

where $h_1, h_2, \ldots, h_s$ are rational integers (D. § 160; B. § 13). Since the equation $F(\theta) = 0$ is irreducible, this results in an equation that holds identically[88] in the variable t of the form

$$A^s + ph_1 A^{s-1} + \cdots + p^s h^s = FG,$$

which implies also the congruence

$$A^s \equiv 0 \pmod{p, F}.$$

Therefore[89] the function A must be divisible modulo p by every prime function that divides F modulo p (C. 5 and 6). Now taking the equation above that is satisfied by the number $A(\theta)$ and multiplying it by $B(\theta)^s$, and recalling that $A(\theta)B(\theta) = pM(\theta)$ we get[90]

$$M(\theta)^s + h_1 M(\theta)^{s-1} B(\theta) + \cdots + h_s B(\theta)^s = 0,$$

and therefore an identity of the form

$$M^s + h_1 M^{s-1} B + h_2 M^{s-2} B^2 + \cdots + h_s B^s = FH.$$

so[91] $M^s \equiv 0 \pmod{p, B}$, which again implies that any prime function dividing B modulo p must also divide M. But we proved above that B is not a constant, so it has at least one prime divisor P, which must then also divide M. Since F is a multiple of B mod p, it must also divide F. But every prime function dividing F must divide A, as we showed above, so P must divide both A and B, which shows P^2 must divide F, since $F \equiv AB \pmod{p}$. So we have shown[92] that there is a prime function contained in M whose square is contained in F, which is what we wanted to prove.

[85] $\deg(B) = \deg(F) - \deg(A) = n - \deg(A) \neq 0$.

[86] Since $F(\theta) = 0$ and $\varphi(\theta)$ is divisible by p, it follows that $A(\theta)$ is divisible by p. In particular, A cannot be a constant mod p. This transfers the assumption that $\varphi(\theta)$ is divisible by p to $A(\theta)$ where $A|F$ in $\mathbb{F}_p[t]$.

[87] The asuumption that $\varphi(\theta)$ is divisible by p leads to the conclusion that $\frac{1}{p}A(\theta)$ is an algebraic integer, so it satisfies a monic equation with integer coefficients. Multiplying by a power of p gives the equation below.

[88] I.e., an equation in $\mathbb{Z}[t]$.

[89] This is the key conclusion. Since A^s is divisible by F in $\mathbb{F}_p[t]$, every irreducible factor of F must also divide A.

[90] After dividing by p^s.

[91] B is a divisor of F in $\mathbb{F}_p[t]$, so every term but the first is divisible by B in $\mathbb{F}_p[t]$.

[92] Factoring out the highest powers of P dividing A and B we get $F \equiv P^e A'B' \pmod{p}$, with $P \nmid A'B'$. By unique factorization in $\mathbb{F}_p[t]$, $A'B'$ is the rest of the factorization of F and $M = \frac{1}{p}(F - P^e A'B')$ is (a possible choice for) the polynomial we are studying. We know that $e \geq 2$ and P divides M. As we observed above, this property is independent of the choice of M, so it follows that no matter how we factor F we will have $F = P^e S - pM$ with P dividing M.

*In a similar way one can easily show that the criterion for the divisibility by p of a number $\varphi(\theta)$ consists in the congruence $\varphi(t) \equiv 0 \mod p, K$, where K is a completely determined divisor of the function F modulo p.

From II and III, the question of whether p divides k reduces to looking at the factorization

$$F = P_1^{e_1} P_2^{e_2} \dots P_m^{e_m} - pM$$

of any function F into prime functions modulo p.[93] In particular, if F is not divisible by the square of any prime function,[94] so that all the exponents $e_1, e_2, \dots, e_m$ are equal to 1,* or when it happens that none of the prime functions whose squares divide F are contained in M, then k is not divisible by p, and Theorem I from §2 applies. But if there is a prime function dividing M whose square also divides F, then k is divisible by p and the second proof of Theorem II shows that the factorization of the ideal $\mathcal{O}p$ into prime factors is *different* from the one determined in Theorem I.[95]

To this result we add the following remark. If the functions R_1, R_2, ..., R_m are congruent to the functions P_1, P_2, ..., P_m, then we have

$$F = R_1^{e_1} R_2^{e_2} \dots R_m^{e_m} - pN$$

and N certainly does not need to be congruent mod p to M. On the other hand, the divisibility of the index k by p is independent of the choice of [lifts of] the divisors mod p, so we must have that the property of M that is key for this result will also hold for N. This can easily be confirmed directly by calculation.[96] If we denote by Q the product of all the prime functions contained in F whose squares are *not* contained in F, one can, by a suitable choice of the functions $R_1, R_2, \dots, R_m$, always arrive[97] at a function N which is relatively prime to Q, but if there is a prime function P that divides M such that P^2 divides F, a calculation shows that then P divides N as well.†[98]

---

[93]All we need to check is there exists an i such that $e_i \geq 2$ and P_i divides M.

[94]This is not the interesting case, since, as Dedekind's footnote points out, if F is not divisible by the square of a polynomial mod p it follows that p does not divide the polynomial discriminant $\Delta = Dk^2$, so of course it does not divide k.

[95]Dedekind doesn't explain why, but we think it might be this. If $F \equiv P^e S \pmod{p}$, P does not divide S, and $\mathfrak{p} = \gcd(p, P(\theta))$, we expect that $\mathfrak{p}^e$ is a divisor of p. But in the second proof of Theorem II we showed that if $p|k$ the number $\eta = S(\theta)P(\theta)^{e-1}$ is divisible by p, but it is not divisible by $\mathfrak{p}^e$. Hence $\mathfrak{p}^e$ does not in fact divide p.

[96]We did this in footnote 72. It is not quite clear that Dedekind's argument works without proving this first, because it seems that M might change in the middle of his proof. In any case, here he notes that it can easily be checked by a direct calculation.

[97]If $F = PB \pmod{p}$ with $P \nmid B$ in $\mathbb{F}_p[t]$, we can always replace P with $P + pC$ where $P \nmid C$ in $\mathbb{F}_p[t]$. That replaces M by $M + CB$, which is not divisible by P in $\mathbb{F}_p[t]$.

[98]The reference in Dedekind's footnote, in which "12" should read "17," is **[127]**.

*This will be the case if and only if the discriminant $\Delta(1, \theta, \theta^2, \dots, \theta^{n-1})$ of the equation $F(\theta) = 0$ is not divisible by p.

†It follows from this that the ideal theory of Zolotareff is limited to the case in which the index k is not divisible by p. At least this seems to follow from the following words, which we can find in the abstract mentioned above (*Jahrbuch über die Fortschritte der Mathematik*, Vol. 6): "To present the theory in its simplest form, the author assumes that $F_1(x)$ is not divisible by any of the functions $V, V_1, V_2 \dots$ If this condition does not hold, one can transform the equation $F(x) = 0$ modulo p so that it does hold. The author reserves the discussion of this transformation for another opportunity." — Since according to my investigations (see §5 of this paper) there exist fields in which the indices of *all* integers θ are divisible by a certain prime number p, it follows that *all* equations $F(\theta) = 0$ have the unfortunate property that impedes the application of Zolotareff's theory. Hence I suppose that there is a misunderstanding in the quoted words from the abstract. It is possible that the author's completion of the theory will be based on considerations similar to those in Selling's theory of ideal numbers (Schlömich's *Zeitschirft*, Vol. 10, p. 12ff.)

§4

In the number domains $\mathcal{O}$ first considered by Kummer, which come from a primitive root of the equation $\theta^m = 1$, the happy circumstance occurs that the powers $1, \theta, \theta^2, \dots, \theta^{n-1}$, with $n = \varphi(m)$, form a basis[99] of the domain $\mathcal{O}$. It follows that the index k of the number θ, on which the entire investigation is based, is $= 1$. I soon realized, however, that in the general investigation of *any* finite field Ω and domain $\mathcal{O}$ containing all the integers in Ω this simple case rarely occurs.[100] I thought for a long time that it was likely that for every given prime number p it might be possible to find an integer θ in the field Ω whose index is not divisible by p. If so, with the help of that θ we could succeed in determining the ideal factors of p. Since all my attempts to prove the existence of such a number θ were unsuccessful, I finally decided, if possible, to prove that this assumption was false. I achieved this goal, as I have already indicated in the *Göttingischen gelehrten Anziegen* of September 20, 1871, through the considerations that form the content of this and the following paragraph.[101]

Let p be a fixed prime number and let[102] $\mathfrak{p}_1, \mathfrak{p}_2, \dots \mathfrak{p}_m$ be all the distinct prime ideals dividing p; we will denote their degrees by $f_1, f_2, \dots, f_m$, so that, for example, $N(\mathfrak{p}_1) = p^{f_1}$. If there exists an integer θ whose index k is not divisible by p, it follows from Theorem I that there exist m polynomials $P_1, P_2, \dots, P_m$ of degrees $f_1, f_2, \dots, f_m$, pairwise *incongruent* modulo p.[103] It is now of the greatest importance for our investigation that this conclusion may be reversed,[104] so that the following theorem holds.[105]

[IV] *Let $f_1, f_2, \dots, f_m$ be the degrees of the distinct prime ideals $\mathfrak{p}_1, \mathfrak{p}_2, \dots, \mathfrak{p}_m$ contained in p. Suppose that modulo p there exist m incongruent prime functions $P_1, P_2, \dots, P_m$ of degrees $f_1, f_2 \dots, f_m$ respectively. Then there exists an integer θ in Ω whose index k is not divisible by p.*

---

[99]If θ is an m-th root of unity, we have $\mathcal{O} = \mathbb{Z}[\theta]$, which is what allows Kummer's general approach to work.

[100]Dedekind says "ausnahmeweise", so "rarely" or "exceptionally." It is unclear how many examples he knew at this point, so this is an impressive insight. One expects that in fact the set of number fields with monogenic rings of integers has density zero. See, for example, [**1**].

[101]This is one of Dedekind's few (and brief) accounts of why he ended up creating ideal theory: the "local" approach that might allow one to reduce everything to "higher congruences" is defeated by the existence of common index divisors. Kronecker [**97**, § 25] makes a similar argument.

[102]Dedekind assumes he knows the factorization of p.

[103]Since different prime ideals dividing p correspond to distinct irreducible polynomials in $\mathbb{F}_p[t]$, such polynomials must exist. Conversely, if we cannot find enough irreducible polynomials of the required degrees, there cannot be any polynomial F whose factorization matches the factorization of p, and so k must be divisible by p no matter which θ is chosen.

[104]So the existence of sufficiently many irreducible polynomials is enough to guarantee the existence of the appropriate θ.

[105]On page 456 of [**72**], Hasse says that "In deriving this criterion, Hensel gave the first demonstration of the power of his new foundation of algebraic number theory." The criterion there comes with an explicit formula for the number of irreducible monic polynomnials of degree f in $\mathbb{F}_p[t]$, but this formula was certainly known to Dedekind as well. Hensel proves the same theorem in [**78**]; see Chapter 6.

Before giving a proof of this theorem, we will make a few general observations[106] that do not depend on all of its hypotheses.

Let $\mathfrak{p}$ be a prime ideal dividing p, of degree f.[107] Then *all* the integers ω of the field Ω satisfy the congruence[108]

$$\omega^{p^f} - \omega \equiv 0 \pmod{\mathfrak{p}}$$

(D. § 163;B. § 26, 3°). Now if t is a variable, the function

$$t^{p^f} - t$$

is congruent mod p to the product of all the incongruent-mod-p prime functions whose degree is a divisor of the number f (C. 19).[109] Among them we can choose *at will* a prime function P whose degree if $= f$; this is always possible because there always exists at least one such function (C. 20).[110] Then

$$t^{p^f} - t \equiv P(t)H(t) \pmod{p},$$

and therefore

$$\omega^{p^f} - \omega \equiv P(\omega)H(\omega) \pmod{p}.$$

Since $\mathfrak{p}$ divides p, we see that for *every* number ω contained in $\mathcal{O}$ we have the congruence

$$P(\omega)H(\omega) \equiv 0 \pmod{\mathfrak{p}}.$$

The number of incongruent roots modulo $\mathfrak{p}$ is therefore exactly $= (\mathcal{O} : \mathfrak{p}) = N(\mathfrak{p}) = p^f$, which is the degree of the congruence. Using the same simple arguments as in rational number theory (D. § 26), one can easily prove that a congruence of degree r modulo a prime ideal $\mathfrak{p}$ can have no more incongruent roots in the number domain $\mathcal{O}$ than the degree r. I will omit the proof for brevity.[111] Therefore in our case the congruence $H(\omega) \equiv 0 \pmod{\mathfrak{p}}$ can have at most $(p^f - f)$ incongruent roots, and it follows that the representatives ω of the f other number classes must satisfy the congruence $P(\omega) \equiv 0 \pmod{\mathfrak{p}}$. For our purposes, however, it is sufficient to know that this congruence has at least one root. Let α be one such root, so that

$$P(\alpha) \equiv 0 \pmod{\mathfrak{p}}.$$

We now consider all the number of the form $\varphi(\alpha)$ and we want to prove[112] that the congruence

$$\varphi(\alpha) \equiv 0 \pmod{\mathfrak{p}}$$

is equivalent to the function congruence

$$\varphi(t) \equiv 0 \pmod{p, P}.$$

---

[106] Making "some observations" is Dedekind's way to prove some lemmas. We will indicate each observation with a footnote.

[107] The first lemma says that if $\mathfrak{p}$ divides p and has degree f, then there exists an irreducible polynomial $P \in \mathbb{F}_p[t]$ of degree f and an element $\alpha \in \mathcal{O}$ such that $P(\alpha) \equiv 0 \pmod{p}$.

[108] Since $\mathcal{O}/\mathfrak{p}$ is a field with p^f elements, every nonzero element has order dividing $p^f - 1$.

[109] The (unique up to isomorphism) field with p^f elements contains all the fields with p^d elements with $d|f$. Again we notice that **[18]** is basically a theory of finite fields.

[110] So P is monic, irreducible mod p, and of degree f, and therefore a divisor of $t^{p^f} - t$.

[111] A polynomial over a field cannot have more roots than its degree.

[112] The second lemma says that an element $\varphi(\alpha) \in \mathbb{Z}[\alpha]$ is divisible by $\mathfrak{p}$ if and only if $\varphi(t)$ is divisible by $P(t)$ in $\mathbb{F}_p[t]$. Essentially, α is a generator of the residue field $\mathcal{O}/\mathfrak{p}$ and its minimal polynomial over $\mathbb{F}_p$ is P.

Indeed, if the latter congruence holds, then also

$$\varphi(t) \equiv P(t)\psi(t) \pmod{p},$$

and so

$$\varphi(\alpha) \equiv P(\alpha)\psi(\alpha) \pmod{p},$$

and since both of the numbers p and $P(\alpha)$ are divisible by $\mathfrak{p}$, we get $\varphi(\alpha) \equiv 0$ mod $\mathfrak{p}$. Conversely, if $\varphi(t)$ is *not* divisible by the prime funciton $P(t)$ then $\varphi(t)$ and $P(t)$ will be relatively prime functions, and if follows that there exist two functions $\varphi_1(t), \varphi_2(t)$ such that the congruence

$$\varphi(t)\varphi_1(t) + P(t)\varphi_1(t) \equiv 1 \pmod{p}$$

holds (C. 5). Then we have

$$\varphi(\alpha)\varphi_1(\alpha) + P(\alpha)\varphi_1(\alpha) \equiv 1 \pmod{p},$$

and if follows that $\varphi(\alpha)$ is *not* $\equiv 0 \pmod{\mathfrak{p}}$. So we have proved the claim above.

In the case[113] that p is divisible by $\mathfrak{p}^2$, we also want to choose the root α of the congruence $P(\alpha) \equiv 0 \pmod{\mathfrak{p}}$ so that the number $P(\alpha)$ is *not* divisible by $\mathfrak{p}^2$. This is always possible: if α is a root of the congruence $P(\alpha) \equiv 0 \pmod{\mathfrak{p}^2}$, then one can choose a number λ that is divisible by $\mathfrak{p}$ but not by $\mathfrak{p}^2$ and set $\alpha' = \alpha + \lambda$. Then[114]

$$\begin{aligned} P(\alpha') &= P(\alpha) + \lambda P'(\alpha) + \lambda^2 P''(\alpha) + \ldots \\ &\equiv \lambda P'(\alpha) \pmod{\mathfrak{p}^2} \end{aligned}$$

and since the derivative function $P'(t)$ has degree $(f-1)$ and is not $\equiv 0 \pmod{p}$, it cannot[115] be $\equiv 0 \pmod{p, P}$, and therefore the number $P'(\alpha)$ above is not divisible by $\mathfrak{p}$. So the number $\lambda P'(\alpha)$, and therefore also the number $P(\alpha')$, is divisible by $\mathfrak{p}$ but not by $\mathfrak{p}^2$. So we have proved the existence of a number α'. Let's remove the accent and thus assume that $P(\alpha)$ is divisible by $\mathfrak{p}$ but not by $\mathfrak{p}^2$. [116]

Let[117] $\mathfrak{p}^e$ be the highest power of $\mathfrak{p}$ contained in p; we want to prove[118] that the numerical congruence

$$\varphi(\alpha) \equiv 0 \pmod{\mathfrak{p}^e}$$

is equivalent to the function congruence

$$\varphi(t) \equiv 0 \pmod{p, P^e}.$$

If the latter holds, then

$$\varphi(t) \equiv P(t)^e\psi(t) \pmod{p},$$

so also

$$\varphi(\alpha) \equiv P(\alpha)^e\psi(\alpha) \pmod{p}.$$

---

[113] The third lemma is that if necessary we can change α to make sure $P(\alpha)$ is not divisible by $\mathfrak{p}^2$.

[114] The editors of **[25]** note here that $P''(\alpha)$ really should be $\frac{P''(\alpha)}{2}$ and similarly for higher terms. They do not bother showing that those denominators will not create trouble for the argument. It would be better to avoid Taylor's theorem and derivatives entirely.

[115] Dedekind knows that "finite fields are perfect" but only in the language of **[18]**.

[116] So now we have an irreducible polynomial P of degree f, an element $\alpha \in \mathcal{O}$ such that $P(\alpha) \equiv 0 \pmod{\mathfrak{p}}$; if $\mathfrak{p}^2$ divides p, we can assume that $P(\alpha) \not\equiv 0 \pmod{\mathfrak{p}^2}$.

[117] The fourth lemma is another translation into polynomial congruences, this time for divisibility by a power of $\mathfrak{p}$.

[118] The proof is identical to that of the second lemma. The key result from **[18]** is that the gcd of two polynomials is a linear combination.

Since both p and $P(\alpha)^e$ are divisible by $\mathfrak{p}^e$, it follows that $\varphi(\alpha) \equiv 0 \pmod{\mathfrak{p}^e}$. Conversely, if the function congruence does *not* hold, the greatest common divisor mod p between $\varphi(t)$ and $P(t)^e$ must be of the form $P(t)^s$ for some $s < e$. So (C. 4) we have polynomials $\varphi_1(t), \varphi_2(t)$ such that

$$\varphi(t)\varphi_1(t) + P(t)^e\varphi_2(t) \equiv P(t)^s \pmod{p}.$$

Since both p and $P(\alpha)^e$ are divisible by $\mathfrak{p}^e$, we get

$$\varphi(\alpha)\varphi_1(\alpha) \equiv P(\alpha)^s \pmod{\mathfrak{p}^e}.$$

Since $s < e$ and $P(\alpha)$ is not divisible by $\mathfrak{p}^2$, it follows that $\varphi(\alpha)$ is *not* $\equiv 0 \pmod{\mathfrak{p}^2}$, and our claim is proved.

One can now apply the results described above[119] to each of the prime ideals $\mathfrak{p}_1, \mathfrak{p}_2, \ldots, \mathfrak{p}_m$. One chooses *arbitrary* prime functions $P_1, P_2, \ldots, P_m$ whose degrees $f_1, f_2, \ldots, f_m$ are the same as the degree of the corresponding prime ideal. As above, one determines as many numbers $\alpha_1, \alpha_2, \ldots, \alpha_m$ such that $P(\alpha_1), P(\alpha_2) \ldots, P(\alpha_m)$ are respectively divisible by $\mathfrak{p}_1, \mathfrak{p}_2, \ldots, \mathfrak{p}_m$ and such that in the case that p is divisible by $\mathfrak{p}_r^2$ the corresponding $P_r(\alpha_r)$ is not divisible by $\mathfrak{p}_r^2$.

Since the ideals $\mathfrak{p}_1, \mathfrak{p}_2, \ldots, \mathfrak{p}_m$ are distinct, their squares are [pairwise] relatively prime, so one can find (D. § 163; B. § 26)[120] a number θ such that

$$\begin{aligned} \theta &\equiv \alpha_1 \pmod{\mathfrak{p}_1^2} \\ \theta &\equiv \alpha_2 \pmod{\mathfrak{p}_2^2} \\ &\cdots\cdots \\ \theta &\equiv \alpha_m \pmod{\mathfrak{p}_m^2} \end{aligned}$$

Then we have

$$\begin{aligned} P_1(\theta) &\equiv P_1(\alpha_1) \pmod{\mathfrak{p}_1^2} \\ P_2(\theta) &\equiv P_2(\alpha_2) \pmod{\mathfrak{p}_2^2} \\ &\cdots\cdots \\ P_m(\theta) &\equiv P_m(\alpha_m) \pmod{\mathfrak{p}_m^2} \end{aligned}$$

It follows that the numbers $P_1(\theta), P_2(\theta), \ldots, P_m(\theta)$ are divisible respectively by $\mathfrak{p}_1, \mathfrak{p}_2, \ldots, \mathfrak{p}_m$, but, in the case when p is divisible by $\mathfrak{p}_r^2$, the number $P_r(\theta)$ is *not* divisible by $\mathfrak{p}_r^2$. The number θ therefore unites in itself all the properties that each of the numbers α_r has with respect to the corresponding prime ideal $\mathfrak{p}_r$.[121] Now[122] let

$$\mathcal{O}p = \mathfrak{p}_1^{e_1}\mathfrak{p}_2^{e_2} \ldots \mathfrak{p}_m^{e_m},$$

so that we have, by taking the norm,

$$n = e_1 f_1 + e_2 f_2 + \cdots + e_m f_m.$$

---

[119]The final lemma shows, using the Chinese Remainder Theorem, that we can find a single θ satisfying the same properties as the α_i. This will, of course, eventually be the θ whose existence is claimed in Theorem IV.

[120]This is the Chinese Remainder Theorem in $\mathcal{O}$; Dedekind does not use that name.

[121]So now we have a single number θ that "works" for all i simultaneously, rather than individual α_i.

[122]We think Dedekind has finished the "general observations" mentioned above, so that the proof of Theorem IV now begins. One could, however, argue that this is one more lemma, and that the real proof begins when he invokes the key assumption in the next paragraph.

A number of the form $\varphi(\theta)$ is divisible by one of the powers $\mathfrak{p}_1^{e_1}, \mathfrak{p}_2^{e_2}, \ldots, \mathfrak{p}_m^{e_m}$ if and only if the corresponding function congruence

$$\varphi(t) \equiv 0 \pmod{p, P_1^{e_1}}$$
$$\varphi(t) \equiv 0 \pmod{p, P_2^{e_2}}$$
$$\ldots\ldots$$
$$\varphi(t) \equiv 0 \pmod{p, P_m^{e_m}}$$

holds. An integer from the field is divisible by p if and only if it is divisible by *each* of the m powers $\mathfrak{p}_1^{e_1}, \mathfrak{p}_2^{e_2}, \ldots, \mathfrak{p}_m^{e_m}$, and therefore a numerical congruence

$$\varphi(\theta) \equiv 0 \pmod{p}$$

is equivalent to the *system* of m function congruences above.

So far we have intentionally put no restriction on the *choice* of the prime functions P_1, P_2, $\ldots$, P_m except that their degrees are respectively those of the prime ideals $\mathfrak{p}_1, \mathfrak{p}_2, \ldots, \mathfrak{p}_m$, so that, for example, if $f_1 = f_2$ nothing stops us from choosing $P_1 = P_2$. We now want to introduce the main assumption[123] of the theorem, namely that we can find m *pairwise incongruent* prime functions of the desired degree, and we will assume that $P_1, P_2, \ldots, P_m$ are such pairwise incongruent prime functions. Then the powers $P_1^{e_1}, P_2^{e_2}, \ldots, P_m^{e_m}$ will be pairwise relatively prime; if we let

$$R = P_1^{e_1} P_2^{e_2} \ldots P_m^{e_m}$$

be their product, then[124] the numerical congruence

$$\varphi(\theta) \equiv 0 \pmod{p}$$

is equivalent the system of m function congruences given above and so (C. 5)[125] is equivalent to the [single] function congruence

$$\varphi(t) \equiv 0 \pmod{p, R}.$$

Notice that the degree of the product R is

$$e_1 f_1 + e_2 f_2 + \cdots + e_m f_m,$$

and so is $= n$. Therefore[126] a number

$$\varphi(\theta) = x_0 + x_1\theta + x_2\theta^2 + \cdots + x_{n-1}\theta^{n-1}$$

can only be divisible by p if $\varphi(t) \equiv 0 \pmod{p}$, i.e., only if all the x_j are divisible by p. It follows (from §1) that the index of θ is *not* divisible by p. So we have proved[127] the Theorem stated above, and we now want to add the following remark.[128]

Since k is not divisible by p, it is also not equal to 0. Thus, the number θ we obtained is the root of an irreducible equation $F(\theta) = 0$ of degree n. Then $F(\theta) \equiv 0$

[123]The key assumption is used here.

[124]First use the equivalence we just proved.

[125]This is the Chinese Remainder Theorem for polynomial congruences.

[126]Since $\varphi(t)$ is a polynomial of degree $n-1$ it can only be divisible by R modulo p if it is zero modulo p.

[127]If there are enough irreducible polynomials modulo p, we can find a number θ whose index is not divisible by p.

[128]The remark is just that the polynomials we have found are exactly the factors (modulo p) of the irreducible polynomial of θ.

(mod p), so the function F must be divisible by R mod p. Since both functions have degree n and have highest coefficient 1, we must have $F \equiv R \pmod{p}$, i.e.

$$F \equiv P_1^{e_1} P_2^{e_2} \dots P_m^{e_m} \pmod{p},$$

and we have now returned to the starting point of our investigation in §2.[129]

§5

Our last investigation has yielded a criterion that answers the question[130] of whether Ω contains an integer θ whose index is not divisible by p. When we have

$$\mathcal{O}p = \mathfrak{p}_1^{e_1} \mathfrak{p}_2^{e_2} \dots \mathfrak{p}_m^{e_m},$$

where $\mathfrak{p}_1, \mathfrak{p}_2, \dots, \mathfrak{p}_m$ are distinct prime ideals whose degrees are, respectively, f_1, f_2, ..., f_m, then the singular case in which the indices of *all* the integers in Ω are divisible by p happens when and only when it is not possible to find m prime functions of degree $f_1, f_2, \dots, f_m$ that are pairwise incongruent mod p. Now we must ask whether the case when there are not enough such polynomials ever does occur. To answer this, we will take the simplest possible approach. The incongruent prime functions *of degree one* are the following:

$$t, t+1, t+2, \dots, t+(p-1).$$

Their number is $= p$. So the singular case above will occur in a field Ω whenever a prime number p factors as the product of $p+1$ distinct prime ideals of degree 1. By the norm computation above, the degree n of such a field must be[131] $= p+1$. If, in order to obtain the simplest case, one takes the smallest prime number $p = 2$, the question arises whether there are *cubic* fields Ω in which the number 2 is divisible by three distinct prime ideals of degree one.[132] In such a field, the indices of *all* algebraic integers will be *even*. This investigation was carried out in full generality the *Göttingischen gelehrten Anzeigen* of September 20, 1871, and led to

[129] To summarize, Dedekind has proved the following: if we know the factorization of p in Ω and we choose *any* list of polynomials of the appropriate degrees, then we can find an element θ that generates Ω over $\mathbb{Q}$ and whose index is not divisible by p. It will not necessarily be a nice generator.

For example, suppose $\Omega = \mathbb{Q}(\sqrt{2})$ and $p = 7$. Then in fact $\mathcal{O} = \mathbb{Z}[\sqrt{2}]$. The principal ideal $\mathcal{O}(7)$ factors as $\mathcal{O}7 = \mathcal{O}(3+\sqrt{2}) \cdot \mathcal{O}(3-\sqrt{2})$. The factors $\mathfrak{p}_1 = \mathcal{O}(3+\sqrt{2})$ and $\mathfrak{p}_2 = \mathcal{O}(3-\sqrt{2})$ are both prime ideals of degree one. Let's deliberately make the "wrong" choice of two distinct polynomials of degree one in $\mathbb{F}_7[t]$: $P_1 = t$, $P_2 = t-1$. We now need to find α_1, α_2 such that $P_i(\alpha_i) \equiv 0 \pmod{\mathfrak{p}_i}$. Clearly we can take $\alpha_1 = 3+\sqrt{2}$ and $\alpha_2 = 4-\sqrt{2}$. Solving $\theta \equiv \alpha_i \pmod{\mathfrak{p}_i^2}$, we get $\theta \equiv 25 + 27\sqrt{2} \pmod{49}$, so let $\theta = 25 + 27\sqrt{2}$. The minimal polynomial for θ is

$$\theta^2 - 50\theta - 833 \equiv \theta^2 - \theta \pmod{7}$$

as expected, and the disciminant of θ is 5832, so the index of $\mathbb{Z}[\theta]$ is $729 = 3^6$, which is not divisible by 7.

[130] More specifically, it gives an answer to the question of whether θ exists if we know the factorization of the principal ideal $\mathcal{O}p$.

[131] More generally, Dedekind's argument shows that if p splits completely in a field of degree $n > p$, then it will be a common index divisor.

[132] So we can find fields with common index divisors if we can solve the following problem: given a prime p and an integer $n > p$, find a field of degree n in which p splits completely. Dedekind chooses $p = 2$ and $n = 3$.

an affirmative answer; here I will be content[133] to give a single example[134] that has already been mentioned there.[135]

Let α be a root of the irreducible polynomial of degree 3

$$F(\alpha) = \alpha^3 - \alpha^2 - 2\alpha - 8 = 0.$$

To find the discriminant,[136] we find the number

$$F'(\alpha) = \delta = -2 - 2\alpha + 3\alpha^2.$$

Then, repeatedly using $F(\alpha) = 0$, we compute the products

$$\delta\alpha = 24 + 4\alpha + \alpha^2$$
$$\delta\alpha^2 = 8 + 26\alpha + 5\alpha^2$$

and via linear elimination[137] of $1, \alpha, \alpha^2$ we find that δ is a root of the equation

$$\begin{vmatrix} -2-\delta & -2 & 3 \\ 24 & 4-\delta & 1 \\ 8 & 26 & 5-\delta \end{vmatrix} = 0$$

that is,

$$\delta^3 - 7\delta^2 - 2012 = 0.$$

Therefore we have

$$\Delta(1, \alpha, \alpha^2) = -N(\delta) = -2012 = -2^2 \cdot 503.$$

Since 503 is a prime number, the only two square divisors of this discriminant are 1 and 4, so the index k of the number α is either 1 or 2. It is therefore the function

$$F(t) = t^3 - t^2 - 2t - 8$$

that we need to investigate modulo $p = 2$. Clearly

$$F = P_1^2 P_2 - 2M \equiv P_1^2 P_2 \pmod{2},$$

where

$$P_1 = t, \quad P_2 = t - 1, \quad M = t + 4.$$

[133] In the *Anzeige* Dedekind described a way to find such fields. His account makes it clear that there are infinitely many examples. By contenting himself here to giving a single example, he may have confused some of his readers. Kronecker, for instance, praised Hensel's thesis for providing infinitely many examples of fields with CIDDs, which in fact Dedekind had already done.

[134] Dedekind does much more than just give the example. Starting from an irreducible polynomial of degree three, Dedekind finds an integral basis, computes the discriminant, finds an explicit factorization of 2, and computes explicitly the products of all the ideals that divide 2. One gets the impression that he wants to show that everything in his theory can be computed explicitly, given enough time and parience.

[135] The editors of **[25]** add a footnote: "The notice [Anzeige] just mentioned contains an explanation of the method by which Dedekind came up with the example discussed here. It also contains another example of a field with a common index divisor, namely a quartic field in which the prime number 2 decomposes into two prime ideals of degree two."

[136] Dedekind uses the theorem that the discriminant of the minimal polynomial for α is equal to $(-1)^{n(n-1)/2} N(F'(\alpha))$.

[137] We would say "using the Cayley-Hamilton theorem."

Since P_1 is a factor of M and P_1^2 is a factor of F modulo 2, it follows[138] (from the second proof of Theorem II in §3) that the number

$$P_1(\alpha)P_2(\alpha) = \alpha(\alpha+1)$$

is divisible by 2, and therefore $k = 2$. That is immediately confirmed by the fact that the number

$$\beta = \frac{1}{2}\alpha(\alpha-1) - 1$$

turns out to be an algebraic integer. In fact, using the equation $F(\alpha) = 0$ we find that

$$\begin{aligned}\alpha^2 &= 2 + \alpha + 2\beta\\ \beta^2 &= -2 + 2\alpha - \beta\\ \alpha\beta &= 4\end{aligned}$$

and so[139]

$$\beta^3 + \beta^2 + 2\beta - 8 = 0.$$

It follows that

$$\begin{aligned}1 &= 1\cdot 1 + 0\cdot\alpha + 0\cdot\beta\\ \alpha &= 0\cdot 1 + 1\cdot\alpha + 0\cdot\beta\\ \alpha^2 &= 2\cdot 1 + 1\cdot\alpha + 2\cdot\beta\end{aligned}$$

and so

$$\Delta(1,\alpha,\alpha^2) = \begin{vmatrix}1 & 0 & 0\\ 0 & 1 & 0\\ 2 & 1 & 2\end{vmatrix}^2 \Delta(1,\alpha,\beta) = 2^2\Delta(1,\alpha,\beta),$$

from which we see that

$$\Delta(1,\alpha,\beta) = -503.$$

Since this number is not divisible by any square (except 1), it is the fundamental number D of our cubic field Ω, and the numbers $1,\alpha,\beta$ are a basis for all the integers ω belonging to the domain $\mathscr{O}$, i.e.,

$$\mathscr{O} = [1,\alpha,\beta]$$

in the notation we have used before.[140] Any algebraic integer in $\mathscr{O}$ can be written in the form

$$\omega = z + x\alpha + y\beta,$$

where z, x, y are arbitrary rational integers.

We now want to use these results to determine the factorization[141] of the number 2. Since

$$\begin{aligned}\alpha^2 &= 2 + \alpha + 2\beta \equiv \alpha \pmod{2}\\ \beta^2 &= -2 + 2\alpha + \beta \equiv \beta \pmod{2}\end{aligned}$$

[138] If we believe the theorems we are done: this is the condition for 2 to be an index divisor for this polynomial, and so $k = 2$. But Dedekind will check that his theorems are true each time.

[139] The minimal polynomial shows that β is an integer.

[140] We have computed the discriminant D and an integral basis of $\mathscr{O}$.

[141] This is the key: Dedekind wants to check that 2 splits completely, but he cannot use Theorem I, so he goes for a direct computation. The first step is to show 2 is unramified.

we get

$$(z + x\alpha + y\beta)^2 \equiv z^2 + x^2\alpha^2 + y^2\beta^2 \equiv z + x\alpha + y\beta \pmod 2,$$

so that *every* $\omega \in \mathcal{O}$ satisfies $\omega^2 - \omega \equiv 0 \pmod 2$. If follows, first, that 2 cannot be divisible by the square of a prime ideal.[142] Indeed, if $\mathcal{O}(2) = \mathfrak{p}^2\mathfrak{q}$, with $\mathfrak{p}$ a prime ideal in $\mathcal{O}$ or even any ideal different from $\mathcal{O}$, then $\mathfrak{pq}$ is not divisible by $\mathcal{O}(2)$, so there exists an element ω such that $\mathfrak{pq}$ divides ω but 2 does not divide ω. Then ω^2 is divisible by $\mathfrak{p}^2\mathfrak{q}^2$ and therefore by 2, and this contradicts[143] the congruence $\omega^2 \equiv \omega \pmod 2$ above. So $\mathcal{O}(2)$ is either a prime ideal or a product of distinct prime ideals.[144] Let $\mathfrak{p}$ be a prime ideal dividing 2. Then we must have $\omega^2 \equiv \omega \pmod{\mathfrak{p}}$ for *every* ω in $\mathcal{O}$. The number of incongruent roots of this congruence is $(\mathcal{O} : \mathfrak{p}) = N(\mathfrak{p})$, but the number of roots cannot be larger than the degree of the congruence, so we get $N(\mathfrak{p}) \leq 2$ and hence $N(\mathfrak{p}) = 2$. So $\mathfrak{p}$ is a prime ideal and is not all of $\mathcal{O}$, since $N(\mathfrak{p}) > 1$. Therefore every prime ideal contained in 2 is of degree one, and it follows, since $N(2) = 2^3 = 8$, that

$$\mathcal{O}(2) = \mathfrak{abc},$$

where $\mathfrak{a}$, $\mathfrak{b}$, $\mathfrak{c}$ are *three* distinct prime ideals of degree one.[145] This shows that the singular case described above does occur, and we will check[146] that indeed the indices of *all* the number ω will be divisible by 2. In fact, if we set[147]

$$\begin{aligned} z' &= z^2 + 2x^2 - 2y^2 + 8xy \\ x' &= x^2 + 2y^2 + 2xz \\ y' &= 2x^2 - y^2 + 2yz \end{aligned}$$

we have

$$\omega^2 = z' + x'\alpha + y'\beta,$$

from which it follows that the index of ω is equal to the determinant

$$\begin{vmatrix} 1 & 0 & 0 \\ z & x & y \\ z' & x' & y' \end{vmatrix} = xy' - x'y = 2x^3 - x^2y - xy^2 - 2y^3,$$

which is always *even*.[148]

In order to complete our example and to confirm the predictions derived from general theory by calculation,[149] we want finally to represent the ideals that appear here in the form of *finite modules with three generators*[150] (D. § 161; B.§ 3),

[142] The next two sentences give a proof.

[143] Indeed, $0 \equiv \omega^2 \not\equiv \omega \pmod 2$.

[144] Now we need to prove 2 is not a prime in $\mathcal{O}$.

[145] We are done. But Dedekind will prove it again.

[146] Again, we already know this must happen, but Dedekind will check it explicitly.

[147] Dedekind is just computing ω^2.

[148] This will later be called computing the "index form," especially in the Kronecker school. Hensel showed in **[79]** that the index form has content 1, i.e., it does not have an integer factor bigger than 1. On the other hand, it may be that the *values* of the index form are always divisible by some prime. Here the form is congruent mod 2 to $(x^2 - x)y - (y^2 - y)x$ and hence is always $0 \pmod 2$. In **[78]**, which we translate in Chapter 6, Hensel proved that p is a common divisor of the values if and only if the form involves $u^p - u$, as here.

[149] So we are going to check everything explicitly.

[150] Dedekind say "dreigliedrigen Moduln," which means something like "triple modules" or "trinomial modules."

i.e., to determine these ideals by finding their bases. These representations are as follows:[151]

$$\mathfrak{a} = [2, \alpha, 1+\beta]$$
$$\mathfrak{b} = [2, 1+\alpha, \beta]$$
$$\mathfrak{c} = [2, \alpha, \beta].$$

The system $\mathfrak{a}$ of all numbers of the form

$$\alpha' = 2z + \alpha z + (1+\beta)y,$$

where z, x, y are arbitrary rational integers, indeed has the fundamental properties of an ideal, namely:

I. The sums and differences of two numbers α' in the system $\mathfrak{a}$ belong to the same system $\mathfrak{a}$.

II. Each product of a number α' from the system $\mathfrak{a}$ and a number ω from the domain $\mathcal{O}$ is still a number from the system $\mathfrak{a}$.

The first property is clear. To prove the second it suffices to check that the product of each of the basis numbers 2, α, $1+\beta$ of $\mathfrak{a}$ by each of the basis numbers 1, α, β of $\mathcal{O}$ belongs to $\mathfrak{a}$. This is clear right away for the five products

$$2.1, \alpha.1, (1+\beta).1, 2.\alpha, 2.\beta = -2 + 2(1+\beta).$$

For the remaining four the same follows from the equations

$$\alpha.\alpha = \alpha + 2(1+\beta), \quad \alpha.\beta = 2.2,$$

$$(1+\beta)\alpha = 2.2 + \alpha, \quad (1+\beta)\beta = -2 + 2\alpha.$$

In the same way one can ckeck that the systems $\mathfrak{b}$ and $\mathfrak{c}$ are ideals.

The *Norm* $N(\mathfrak{m})$ of an ideal $\mathfrak{m}$ is the number $(\mathcal{O} : \mathfrak{m})$ of numbers that are incongruent mod $\mathfrak{m}$ (D. § 163; B. § 20), which is equal to the determinant of the expressions that give the basis numbers of $\mathfrak{m}$ as linear combinations of the basis numbers of $\mathcal{O}$ (D. § 161; B. § 4, 4°). So, for example,

$$N(\mathfrak{a}) = \begin{vmatrix} 2 & 0 & 0 \\ 0 & 1 & 0 \\ 1 & 0 & 1 \end{vmatrix} = 2,$$

and in the same way

$$N(\mathfrak{b}) = N(\mathfrak{c}) = 2.$$

When, however, the norm of an ideal is a prime number, that ideal must necessarily be a prime ideal,[152] since in general we have $N(\mathfrak{a}_1\mathfrak{a}_2) = N(\mathfrak{a}_1)N(\mathfrak{a}_2)$. Therefore $\mathfrak{a}$, $\mathfrak{b}$, $\mathfrak{c}$ are prime ideals. Further, they are pairwise distinct, since the number β belongs to both $\mathfrak{b}$ and $\mathfrak{c}$ but not to $\mathfrak{a}$, and the number α, which is contained in $\mathfrak{c}$, is not contained in $\mathfrak{b}$. The number 2 is contained in all three ideals and so must also be contained in the product $\mathfrak{abc}$, so $\mathcal{O}(2) = \mathfrak{mabc}$, where $\mathfrak{m}$ is some ideal. But computing the norm we get

$$N(2) = 8 = N(\mathfrak{m})N(\mathfrak{a})N(\mathfrak{b})N(\mathfrak{c}) = 8N(\mathfrak{m}),$$

[151] Dedekind doesn't explain how he computed these (it's easy enough), but he will check that these modules are indeed ideals and that their product is 2.

[152] This confirms that the three factors are prime ideals.

therefore $N(\mathfrak{m}) = 1$, so $\mathfrak{m} = \mathcal{O}$ and $\mathcal{O}(2) = \mathfrak{abc}$.[153] But we also want to check this result, which follows from general theorems, by a direct computation, i.e., through the actual *multiplication* of the ideals (D. § 165; B. § 12).[154]

By the *product* $\mathfrak{ab}$ of two ideals we understand the system of all products $\alpha'\beta'$ and all sums of such products $\alpha'\beta'$, where α', β' are any numbers belonging respectively to the ideals $\mathfrak{a}$, $\mathfrak{b}$ (D. §163; B. § 22). Such a product is therefore first a finite module whose basis numbers are all the products of each basis number of $\mathfrak{a}$ by each basis number of $\mathfrak{b}$. In our case, then, $\mathfrak{ab}$ is the finite module whose basis numbers are the nine products

$$2.2 = 4, \quad 2(1+\alpha) = 2 + 2\alpha, \quad 2.\beta = 2\beta,$$

$$\alpha.3 = 2\alpha, \alpha(1+\alpha) = 2 + 2\alpha + 2\beta, \quad \alpha\beta = 4,$$

$$(1+\beta).2 = 2 + 2\beta, \quad (1+\beta)(1+\alpha) = 5 + \alpha + \beta, \quad (1+\beta)\beta = -2 + 2\alpha.$$

Of those nine number only three are mutually *independent* (D. § 159; B. § 4), so by the method I have described in detail (B. § 4,6°), we can reduce this module with nine generators[155] to one with three generators.[156] Doing this very simple and easy calculation one gets the six following equations:

$$\mathfrak{a}^2 = [4, \alpha, 3+\beta]; \qquad \mathfrak{bc} = [2, 2\alpha, \beta]$$

$$\mathfrak{b}^2 = [4, 1+\alpha, \beta]; \qquad \mathfrak{ca} = [2, \alpha, 2\beta]$$

$$\mathfrak{c}^2 = [4, 2+\alpha, 2+\beta]; \qquad \mathfrak{ab} = [2, 2\alpha, 1+\alpha+\beta]$$

We now proceed in the same way. Multiplying each of those by $\mathfrak{a}$, $\mathfrak{b}$, $\mathfrak{c}$ using the same method, we obtain the following ten principal ideals:

$$\begin{aligned}
\mathfrak{abc} &= [2,\ 2\alpha,\ 2\beta] = \mathcal{O}(2)\\
\mathfrak{a}^2\mathfrak{c} &= [4,\ \alpha,\ 2+2\beta] = \mathcal{O}\alpha\\
\mathfrak{b}^2\mathfrak{c} &= [4,\ 2+2\alpha,\ \beta] = \mathcal{O}\beta\\
\mathfrak{ac}^2 &= [4,\ 2+\alpha,\ 2\beta] = \mathcal{O}(\alpha-2)\\
\mathfrak{bc}^2 &= [4,\ 2\alpha,\ 2+\beta] = \mathcal{O}(2-\beta)\\
\mathfrak{a}^2\mathfrak{b} &= [4,\ 2\alpha,\ 3+\alpha+\beta] = \mathcal{O}(3+\alpha+\beta)\\
\mathfrak{ab}^2 &= [4,\ 2+2\alpha,\ 1+\alpha+\beta] = \mathcal{O}(1+\alpha+\beta)\\
\mathfrak{a}^3 &= [8,\ 4+\alpha,\ 3+\beta] = \mathcal{O}(3+2\alpha+\beta)\\
\mathfrak{b}^3 &= [8,\ 1+\alpha,\ 4+\beta] = \mathcal{O}(1+\alpha)\\
\mathfrak{c}^3 &= [8,\ 2+\alpha,\ 2+\beta] = \mathcal{O}(\alpha+\beta-4)
\end{aligned}$$

The ten numbers μ to which these principal ideals $\mathcal{O}\mu = [\mu,\ \alpha\mu,\ \beta\mu]$ correspond are connected to each other by the following easily checked relations:

$$\alpha(\alpha-2)(1+\alpha) = 2^3; \qquad \alpha\beta = (\alpha-1)(1+\alpha+\beta_2^2$$

[153] We have proved the factorization a second time.

[154] Dedekind's approach initially distinguished between two notions of divisibility. On the one hand, an ideal $\mathfrak{a}$ divides $\mathfrak{b}$ if $\mathfrak{b} \subset \mathfrak{a}$. On the other, $\mathfrak{a}$ divides $\mathfrak{b}$ if there exists an ideal $\mathfrak{c}$ such that $\mathfrak{b} = \mathfrak{ac}$. In **[26]** he says that proving these two notions are equivalent is "the main difficulty of the theory," which he has overcome. Here he has established the factorization of 2 using the first point of view. Now he will check it from the second point of view.

[155] He says "neungliedrigen Modul."

[156] "Dreigliedrigen."

$$(\alpha-2)(3+\alpha+\beta)=2\alpha; \qquad \alpha(2-\beta)=2(\alpha-2)$$
$$(\alpha-2)(3+2\alpha+\beta)=\alpha^2; \qquad \alpha(\alpha+\beta-4)=(\alpha-2)^2$$

This example, to which one add many others, makes it clear that there are fields Ω in which the indices of *all* integers are divisible by the same prime number p. This result is not a welcome one in some respects. Indeed, there are many theorems in the theory of ideals that would be easy to prove via the theory of higher congruences were in not for the fact that Theorem I in §2 requires the assumption that the index k of the integer θ not be divisible by p. We have now seen, however, that in many cases this hypothesis cannot be satisfied no matter which number θ we choose, and it follows that the approach suggested by that Theorem will not work in full generality. For example, I mention the following important theorem which I also used in the *Göttingischen gelehrten Anzeigen* of 20 Spetember 1971:

The fundamental number D of a field Ω is the product of those and only those rational prime numbers p that are divisible by the square of a prime ideal in that field.

If there is an integer in Ω whose index is not divisible by the prime number p, the truth of this result clearly follows very easily[157] from §2. But this obviously does not lead to a proof of the general theorem, and it was only after several unsuccessful attempts that I succeeded in finding a general proof. I must, however, reserve the detailed development of this subject, in which the theorem itself will be considerably generalized, for another occasion.[158]

4.4. Notes by Öystein Ore

In Dedekind's collected works, this paper is followed by some two pages of "Erläuterungen zur vorstehenden Abhandlung" signed by Öysten Ore, one of the three editors. We present only some highlights of what Ore has to say. Quotations are from [**25**, pp. 230–232].

> The problem of generalizing the Kummer theory of ideals in cyclotomic fields to general fields leads naturally to a definition of ideals by means of higher congruences. Selling (*Zeitschr. f. Math n. Phys.*, vol 10, pp. 17–47 (1865)) took this path and succeeded in obtaining an exception-free theory of ideals in Galois fields by using Galois imaginaries and other auxiliary fields. The prime ideal decomposition of a prime number p is derived from the decomposition of the defining equation (mod. p^a) in these auxiliary fields. A proof of the invariance of these ideals, i.e., of their independence of the chosen defining equation, is not provided.

The Selling article is the same one cited by Dedekind, [**127**].

[157] If p divides D, then it divides $d(\theta)$ for any θ, which means that the corresponding polynomial $F(x)$ has a double root modulo p and hence is divisible by the square of an irreducible polynomial. If we assume there is a θ whose index is not divisible by p, we can use Theorem I, which tells us that p is divisible by the square of a prime ideal. Conversely, if p is divisible by the square of a prime ideal, $F(\theta)$ is divisible by the square of an irreducible polynomial mod p, and so $p|d(\theta)$. Since $d(\theta)=k^2D$ and $p\nmid k$ then $p|D$.

[158] That would be [**24**].

"As can be seen from the introduction," Ore says, Dedekind had tried this method as well,[159] but then abandoned it in favor of an abstract theory of ideals as presented in the second edition of Dirichlet's *Zahlentheorie.* Ore points out that this form of the theory does not give us, however, an explicit way to determine the factorization of given numbers in the field. Theorem I solves that problem for primes that do not divide the index, but the existence of common index divisors (or common inessential discriminant divisors) blocks that path in general.

In the next few paragraphs, Ore focuses on Zolotarev, noting that in his first paper on algebraic integers Zolotarev had used something like Dedekind's factorization theorem to define ideal prime factors. Of course, this was limited to primes that were not index divisors. Ore says that Zolotarev's second paper **[144]** solved the problem, but to do that had to abandon the approach based on higher congruences. Ore goes on to explain Zolotarev's "semi-local" approach; there are good expositions in **[132]** and **[49]**. Ore says he will not get into all the alternative ways to lay the foundations.

Ore then turns to Kronecker's theory

> Kronecker's theory of forms (*Journ. f. Math.*, Vol. 92, pp. 1–122 (1882)) gives a theoretically very simple determination[160] of the prime ideal decomposition of rational prime numbers. As was first shown in full generality by Hensel (*Journ. f. Math.*, Vol. 113, pp. 61-83 (1894)), this theory contains a complete analogue to Dedekind's theorem for all prime numbers.

The reference is to **[79]**, where Hensel proves a theorem first stated by Kronecker (see the next chapter) to the effect that one can determine the factorization of a rational prime p by factoring the "Fundamentalgleichung" modulo p.

> However, this solution to the problem does not provide any information about the relationship between the properties of the equations of the field and the prime ideal decomposition, as is the case with Dedekind's theorem. In the p-adic theory of algebraic numbers founded by Hensel, this gap is partially filled by showing that the decomposition of the defining equation into irreducible p-adic factors corresponds to the decomposition of p into powers of prime ideals. For the complete determination of the prime ideal decomposition, however, one must still use Kronecker's theory. (See K. Hensel: *Theorie der algebraischen Zahlen I*, Leipzig 1908.)[161]

It is indeed the case that in Hensel's book there is no clean statement of the theorem that the p-adic factorization of the defining equation gives enough information to determine the factorization of p. It seems to have been Ore himself who added the final step.

> It can be shown, however, that the difficulties of Dedekind's theory can be completely eliminated if, instead of congruences (mod

[159] This is even clearer from the discussion in the beginning of §4, where Dedekind says he thought for a long time that this would be possible.

[160] eine theoretisch besonders einfache Bestimmung. The point seems to be that Kronecker's theory yields a constructive result on prime factorization (just factor the fundamental equation). This is "theoretically simple" but often not feasible in practice.

[161] The reference is to Hensel's **[82]**; there was never a Volume II.

> p) one always considers congruences (mod p^α) where $\alpha > \delta$ if the discriminant of the corresponding equation is exactly divisible by p^δ. The corresponding irreducible factors are then not determined (mod p^α), but rather (mod $p^{\alpha-\delta}$). The common index divisors then completely lose their exceptional character and one obtains a clear correspondence between prime ideal decomposition and factors of the equation (O. Ore, *Math. Ann.*, Vol. 96, pp. 315–352 (1926) and Vol. 97, pp. 569–598 (1927)). Furthermore, Dedekind's representation of the prime ideals in the form $\beta = (p, \varphi(\theta))$ can be recovered by a method that shows great similarity to the determination of the series development of algebraic functions (O. Ore, *Math. Ann.*, Vol. 99, pp. 84–117 (1928)).

Ore does not express his theorem in p-adic terms, but his factorization modulo p^α translates, via Hensel's Lemma, to a p-adic factorization. A modern account of Ore's result can be found in [**16**, Section 6.2.1].

Since the theorem in §4 gives a criterion for the existence of common index divisors, Ore goes on to discuss other results in this direction. He notes first that Hensel gave a different criterion in terms of the index form in [**78**] (see Chapter 6 for a translation). He also mentions that in the same paper Hensel also proved Kronecker's conjecture that if a number field K has common index divisors then there exists an extension field for which the values of the index form do not have a common divisor. (We discuss this strange theorem in Chapter 6.)

Ore's final note emphasizes that Dedekind's result implies that common index divisors are a "small primes" phenomenon. Hensel's criterion, he says, implies[162] that if K is a number field of degree n for which p is a common index divisor then $p < \frac{1}{2}n(n-1)$. He then notes several improvements on this estimate: E. von Żyliński (which Ore spells "Zylinksky") proved [**145**], using Dedekind's criterion, that in fact $p < n$. M. Bauer showed [**8**] that if $p < n$ there always exists a field of degree n for which p is a common index divisor. (Żyliński's very short paper actually appeared after Bauer's; Bauer seems to have thought that the bound $p < n$ was known.) Ore notes that Bauer's result also follows from general theorems of Hasse [**70**] showing the existence of fields in which p has prescribed factorization.

4.5. Promises Kept, but Mysteries Remain

Richard Dedekind's 1878 paper "On the Relationship between the Theory of Ideals and the Theory of Higher Congruences" fulfilled most of the promises made in the 1871 Anzeige. It included the first proof of Dedekind's factorization theorem, a congruential condition for when a prime divides the index $(\mathcal{O} : \mathbb{Z}[\theta])$, and a necessary and sufficient condition, in terms of the factorization of a rational prime p, for p to be a common inessential discriminant divisor. In a sense, the problem of CIDDs was settled.

The first of these results, Dedekind's factorization theorem, has become a standard result in algebraic number theory. The others, however, are now less well known. This seems largely the result of reading Dedekind's criterion of CIDDs as showing that they are entirely a "small primes" phenomenon and therefore not that interesting. This impression was confirmed, of course, by the results of Zylinsky and

[162] See below, footnote 124 in Section 6.4.

Brauer mentioned by Ore in his notes. The criterion appears (and is attributed to Hasse's advisor Kurt Hensel) in Hasse's big treatise on *Number Theory* [**72**, p. 456], but the issue of CIDDs is for the most part no longer mentioned.

From Dedekind's point of view, then, the story would have been this. Attempting to base his theory of algebraic integers on higher congruences, he discovered "after a long time" that for finitely many primes that way of proceeding was obstructed. Rather than accepting a theory that treated those primes separately, he looked for an approach that allowed a unified treatment, and found that in his theory of ideals. Once that was in place, he then used the existence of CIDDs to justify the need for ideal theory. Finally, he used his ideal theory to determine exactly when a prime was a common inessential discriminant divisor. CIDDs went from obstacle to an argument for the new theory and then to a demonstration of the effectiveness of that theory.

From another point of view, however, the issue was not closed. Suppose our goal is to explicitly determine the factorization of a rational prime p in the ring of algebraic integers $\mathscr{O}$ of a number field K. How can we proceed?

(a) First we find an algebraic integer α which is a field generator, so that $K = \mathbb{Q}(\alpha)$. We consider its minimal polynomial f.
(b) Looking at the factorization of the minimal polynomial modulo p, we can use Dedekind's Theorems II and III to decide whether p divides the index $(\mathscr{O} : \mathbb{Z}[\alpha])$.
(c) If p does not divide the index, Dedekind's Theorem I allows us to determine its factorization in $\mathscr{O}$ from the factorization modulo p of the irreducible polynomial.
(d) Suppose, however, we find that p does divide the index $(\mathscr{O}, \mathbb{Z}[\alpha])$. Then we cannot apply Theorem I. (In fact, we know it will give the wrong answer.) We might try to find a different generator, but there is no guarantee that we can find one, because p might be a CIDD.
(e) *If we already knew the factorization of* p, we could use Dedekind's Theorem IV to decide whether p is a CIDD. But finding the factorization was our goal!

Hensel, in 1894, was able to remove the dependence on knowing the factorization of p. As we will see in Chapter 6, he realized that the data we really need are the number of divisors and their degrees, and found a way to determine those without knowing the complete factorization.

Hensel was a student of Leopold Kronecker. He spent roughly the first decade of his mathematical life focusing on Kronecker's version of algebraic number theory, providing proofs for some results Kronecker had announced and extending the theory. It is necessary, then, to take a quick look at what Kronecker had to say before considering Hensel's work in detail. That is the goal of the next chapter.

CHAPTER 5

Kronecker on CIDDs

The next significant work on common inessential discriminant divisors came in Kurt Hensel's thesis and subsequent papers and especially in the 1894 paper we translate in the next chapter. To set the stage for that, we must get to know Hensel's teacher, Leopold Kronecker. The key text is Kronecker's 1882 *Grundzüge einer arithmetischen Theorie der algebraischen Grössen* (Foundations of an Arithmetical Theory of Algebraic Quantities) [**97**]. As the title indicates, Kronecker's theory deals with all kinds of "algebraic quantities," which includes both algebraic integers and algebraic functions in one or more variables.

Kronecker is notoriously hard to read. In [**129**, art. 139], Smith complains that "Unfortunately, in the brief notices which M. Kronecker has given of his investigations, his methods are indicated only in a very general manner..." Similarly, Gösta Mittag-Leffler explained, in a letter[1] written while visiting Berlin, that "Weierstrass and Kronecker both have the unusual tendency, for Germany, of avoiding publication as much a possible. Weierstrass, as is known, publishes nothing at all, and Kronecker only results without proofs." This is certainly true of the *Grundzüge*, in which many arguments are only sketched out.

Kronecker's most detailed discussion of common inessential discriminant divisors in the number field case occurs in section 25 of his *Grundzüge*. In this chapter, we provide a translation of that passage, which laid out a program for the early work of his student Kurt Hensel. We contextualize the passage by giving a short account of algebraic number theory from Kronecker's point of view. This should also help our readers understand Hensel's work, which we discuss in the next chapter.

5.1. Kronecker's *Grundzüge*

E. E. Kummer, writing in 1859 [**103**, p. 57], reports that Kronecker had developed a theory of general complex integers and their connection with the theory of decomposable forms of all degrees "fully and with great simplicity."[2] Whether the

---

[1]We quote from Rowe's translation in [**121**, pp. 43–44]; see also [**137**, p. 5]. Both translate from [**55**, p. 54], which we have not seen; Vergnerie also gives the text in German, translated from Swedish, found in that source.

[2]Ich kann in Betreff dieser, so wie überhaupt der allgemeinen Sätze, welche allen Theorieen complexer Zahlen gemein sind auch auf eine Arbeit von Hrn. Kronecker verweisen, welche nächstens erscheinen wird, in welcher die Theorie der allgemeinsten complexen Zahlen, in ihrer Verbindung mit der Theorie der zerlegbaren Formen aller Grade, vollständig und in grofsartiger Einfachheit entwickelt ist. (In this regard, as well as with respect to the general theorems which are common to all theories of complex numbers, I can also refer to a work by Mr. Kronecker, which will be published shortly, in which the theory of the most general complex numbers, in its connection with the theory of the composite forms of all degrees, is developed fully and with great simplicity.)

theory was "complete" or not at that point[3] and despite Kummer's claim that it was about to be published, nothing appeared until much later. When Kronecker's theory was finally published, algebraic numbers were a special case of a far more general theory of algebraic quantities which had lost whatever "great simplicity" it originally had. The *Grundzüge einer arithmetischen Theorie der algebraischen Grössen* **[97]** appeared in 1882 in a special volume in honor of Kummer. It was reprinted in Crelle's Journal the same year.

In the same Volume 92 of Crelle's Journal appeared a paper by Dedekind and Weber on the theory of algebraic functions in one variable **[27]**. Their approach pursued an analogy[4] between algebraic numbers and algebraic functions. Since Kronecker's general theory included both this and algebraic numbers as special cases, it may be that the Dedekind-Weber paper was what spurred him to finally publish his ideas.[5] It is clear, in any case, that Kronecker delayed the publication of the paper by Dedekind and Weber in order to make sure his paper would appear first.[6]

In his lecture at the 1950 ICM **[140]**, André Weil noted that

> Kronecker's once famous *Grundzüge* are either forgotten, or are thought of merely as presenting an inferior (and less pure) method for achieving part of the same results, viz., the foundation of ideal-theory and of the theory of algebraic number-fields.

H. M. Edwards made similar points in many articles (for example, **[47]**): it is a mistake to read the *Grundzüge* merely as another way to do algebraic number theory. Rather, Kronecker aimed at a general theory of algebraic quantities. In modern terms, he was interested in the birational geometry of algebraic varieties of all dimensions; from this point of view algebraic numbers are the case of dimension zero. Kronecker's main interest was probably in the more general (and more difficult) cases. This can be seen, for example, in sections 7 and 8. Section 7 discusses the discriminant in the "special cases" (besondere Fälle) in which an integral basis exists, while section 8 treats the general case. The "special cases" are exactly algebraic numbers and algebraic functions in one variable, i.e., the cases considered by Dedekind **[21, 22]** and Dedekind-Weber **[27]**.

The desire to start from fields with many variables and to allow many polynomial equations makes Kronecker's theory considerably harder. If one merely wants algebraic numbers, much of the complexity can be avoided. In modern terms, both $\mathbb{Z}$ and $\mathbb{Q}[z]$ are principal ideal domains, but Kronecker wants to allow the base ring $\mathbb{Q}[x_1, x_2, \dots, x_n]$, which is not a PID. A lot of Kronecker's text is devoted to making sure the theory works in full generality.

A useful guide for reading the *Grundzüge* is "La théorie arithmétique de Kronecker," a chapter in Erwan Penchèvre's history of elimination theory **[116**, p. 423–474**]**. It goes section-by-section through the text and gives a modern interpretation of what Kronecker is doing. We give here a short account of some features of Kronecker's approach.

---

[3] It is possible that Kummer would have considered a theory that excluded the prime divisors of the discriminant to be complete. See Section 2.3.

[4] See **[69]** for an analysis of the role of analogy in the Dedekind-Weber paper.

[5] Clearly the publication of Dedekind's Supplement X **[21]** in 1871 did *not* provide sufficient motivation!

[6] Both papers appear in Volume 92 of *Crelle*, but Kronecker's is in "Heft 1," while Dedekind-Weber is in "Heft 3."

Kronecker himself had proved in [**100**] that for algebraic functions in one variable[7] CIDDs do not occur; his definition of a discriminant in the general case of Section 8 makes CIDDs impossible. Since we are interested in the problem of CIDDs, we focus almost exclusively on the special case of algebraic numbers.

5.2. Kronecker's Point of View

A complete account of the contents of Kronecker's *Grundzüge* would require an entire book. Instead, we give a brief overview with the goal of introducing Kronecker's terminology and basic method.

In the *Grundzüge*, Kronecker worked over a base field which he called the domain of rationality ("Rationalitäts-Bereich"), which could be any field but was most often a finitely-generated purely transcendental extension of $\mathbb{Q}$, allowing $\mathbb{Q}$ itself as the case of zero variables (which is of course the case of algebraic number theory). Quantities that are algebraic over the base field are defined to be of the same genus ("Gattung") if each is a rational function of the other. All rational functions of an algebraic quantity of a given genus form a genus domain ("Gattungs-Bereich"), which we would describe as the field generated by that quantity. Kronecker also introduced the notion of a species ("Art"), which in our language is a subring of the genus domain generated by finitely many elements that are integral over the base. One of the key theorems is that the union of all the species in a genus domain is itself a species; in modern terms, the set of all elements of a genus domain that are integral over the base form a finitely generated ring. When we work over $\mathbb{Q}$, this is the ring of all algebraic integers in the field. The modern language follows Dedekind, referring to a species as an "order" and to Kronecker's principal species ("Haupt-Art") as the "maximal order."

One of the key ideas in Kronecker's approach was to introduce new variables[8] so that he could work with what we might call "generic elements." So, for example, the greatest common divisor of two algebraic integers α and β could be represented by $u\alpha + v\beta$, where u and v are indeterminates. In Dedekind's theory, this would be the ideal generated by α and β, which consists exactly of all possible values of $u\alpha + v\beta$ as u and v take on values in some ring. In Kronecker, u and v are just variables.

Just as an ideal can have different generating sets, different generic elements might represent the same divisor. Hence, Kronecker needs some kind of equivalence condition. What he would like to do is to say that two polynomials in several variables are equivalent (in his language, "two forms define the same divisor") if they differ by a factor which is a polynomial whose coefficients have no common divisor, which Kronecker calls a "primitive form." This doesn't quite make sense for polynomials whose coefficients are algebraic integers, since greatest common divisors may not exist, so the actual definition requires taking norms and considering the resulting polynomials with integer coefficients.

In [**116**, p. 451ff], Penchèvre outlines a modern reading of this equivalence. When the base ring is $\mathbb{Z}$ (the case of algebraic numbers), one would consider the

[7] In characteristic zero, of course; Kronecker did not consider algebraic function fields in characteristic p, for which CIDDs can indeed occur.

[8] In the *Grundzüge*, Kronecker refers to the resulting polynomials as "forms" even though they need not be homogeneous.

ring $R = \mathbb{Z}[u_1, u_2, \dots]$ with countably many indeterminates and let S be the multiplicative subset of primitive forms, i.e., polynomials with content 1. Then we work with the localization $S^{-1}R$. Given a ring of algebraic integers $\mathcal{O}$, Kronecker's divisors would be elements of $\mathcal{O} \otimes S^{-1}R$. This ring is integral over $S^{-1}R$ and has unique factorization, so we get unique factorization of elements of $\mathcal{O}$ as products of divisors.

A different modern interpretation of the *Grunzüge* can be found in H. M. Edwards's **[43]**. Since Edwards is sympathetic to Kronecker's philosophy of mathematics, his account is much closer to the original. His book *Divisor Theory* **[46]** is an attempt to present a modern version of the theory.

5.3. What is in Section 25?

Kronecker mentions inessential discriminant divisors at several points. In the final section of the *Grundzüge*, he introduces the fundamental equation and its discriminant, which leads naturally to a discussion of CIDDs. We provide below a translation of a small portion of this section, in which Kronecker discusses what his theory looks like in the special case when the base ring is $\mathbb{Z}$, i.e., in the case of algebraic number theory. As noted above, this is the only case in which CIDDs occur.

To make it easier to read the portion we translate, we attempt here a quick summary of Section 25 of the *Grundzüge*. Kronecker is working in full generality as usual, but our summary mostly restricts itself to the case of algebraic numbers.[9] We do not use Kronecker's notation, but we keep enough of Kronecker's words to highlight that we are not doing modern "structural" mathematics. We gloss or replace Kronecker's terms with the modern terms where it makes things clearer. In the footnotes, we show how the theory looks in a specific case, using the same example given by Dedekind (see Section 4.3, §5).

We suppose we are given a genus domain $\mathbb{Q}(\alpha)$. Let

$$\omega_1, \omega_2, \dots, \omega_n$$

be a fundamental series (i.e., an integral basis),[10] so that any element of the principal species (i.e., the ring of all algebraic integers in our genus domain) is an integral linear combination of the ω_i.

Each of the ω_i is, of course, a polynomial in α with rational[11] coefficients. If we replace α in that polynomial by another root of its minimal polynomial, we get a conjugate $\omega_i^{(j)}$ of the ω_i.[12] The *discriminant* of the genus domain is then the square of the determinant of the matrix $[\omega_i^{(j)}]$. When we are working over $\mathbb{Z}$, the discriminant is an integer as well.[13]

---

[9] Kronecker knows that the ring of integers (principal species) in a genus domain of degree n can be expressed as linear combinations of $n+m$ generators, where $m \geq 0$. The description we give essentially assumes $m = 0$, i.e., the generators are actually an integral basis. This is Kronecker's "special case" in section 7.

[10] For example, let α be a root of the irreducible polynomial $x^3 - x^2 - 2x - 8$. Then we can take $\omega_1 = 1$, $\omega_2 = \alpha$, $\omega_3 = 4/\alpha = -1 - \frac{1}{2}\alpha + \frac{1}{2}\alpha^2$.

[11] Not necessarily integral, since $\mathbb{Z}[\alpha]$ may not be the maximal order.

[12] Since the other roots do not necessarily belong to $\mathbb{Q}(\alpha)$, we are implicitly working in the normal closure.

[13] This definition is identical to Dedekind's, but Dedekind's computation does not use the definition directly. The discriminant of $\mathbb{Q}(\alpha)$ is -503.

Next Kronecker considered the generic element

$$\xi = x_1\omega_1 + x_2\omega_2 + \dots x_n\omega_n$$

that "represents the species" [**97**, §25, p. 372]. Here we are to think of $x_1, x_2, \dots, x_n$ as indeterminates.[14] Operating coefficient-by-coefficient, we let $\xi = \xi^{(1)}, \xi^{(2)}, \dots, \xi^{(n)}$ be the conjugates of ξ. We can then make the "fundamental equation"

$$F(x) = (x - \xi^{(1)})(x - \xi^{(2)}) \cdots (x - \xi^{(n)}).$$

This is an irreducible polynomial in $n+1$ variables with integer coefficients,[15] but Kronecker and Hensel were more likely to think of it as a polynomial in x with coefficients in $\mathbb{Z}[x_1, x_2, \dots, x_n]$. Notice that if we specialize the x_i to integer values ξ will specialize to an algebraic integer. Since the ω_i are an integral basis, we can get any given algebraic integer in our genus domain this way; choosing the same values in the fundamental equation gives the monic polynomial of degree n in $\mathbb{Z}[x]$ that has our element as a root.[16] Since the fundamental equation is a polynomial in x, we can consider its discriminant, which Kronecker just calls "the discriminant of the fundamental equation." This[17] is, of course, a polynomial in $\mathbb{Z}[x_1, x_2, \dots, x_n]$.

So now we have two discriminants: the discriminant D of the genus domain, which is an integer, and the discriminant $\mathfrak{D}$ of the fundamental equation, which is a polynomial in n variables. If we give integer values to the variables x_i appearing in $\mathfrak{D}$, we get the discriminant of the minimal polynomial of the corresponding algebraic integer, which as we know is always divisible by D.

Next comes a fundamental claim: Kronecker says that as divisors D and $\mathfrak{D}$ are absolutely equivalent. Since $\mathfrak{D}$ has integer coefficients, this boils down to saying that the content of $\mathfrak{D}$ is D and the coefficients of remaining factor, which is the square of the index form, have no common factor.[18] Kronecker and Hensel typically expressed this as "D is the numerical factor of $\mathfrak{D}$", i.e., $\mathfrak{D} = D\,\Delta(x_1, x_2, \dots, x_n)^2$, where the coefficients of Δ have no common factor.

Kronecker's next move is to consider the reduction modulo p of the fundamental equation. Since the fundamental equation might be reducible modulo p, Kronecker considers its factorization

$$F(x) \equiv f_1(x)^{m_1} f_2(x)^{m_2} \dots \pmod{p},$$

[14]The generic element, or fundamental form, of our $\mathbb{Q}(\alpha)$ is $x_1 + \alpha x_2 + \frac{4}{\alpha}x_3$. For the conjugates we just replace α by a different root of our polynomial.

[15]In our example, we get

$$\begin{aligned} F(x) &= x^3 - (3x_1 + x_2 - x_3)x^2 + (3x_1^2 + 2x_1x_2 - 2x_2^2 - 2x_1x_3 - 13x_2x_3 + 2x_3^2)x \\ &\quad - x_1^3 - x_1^2x_2 + 2x_1x_2^2 - 8x_2^3 + x_1^2x_3 + 13x_1x_2x_3 + 6x_2^2x_3 - 2x_1x_3^2 - 10x_2x_3^2 - 8x_3^3. \end{aligned}$$

Probably the easiest way to compute it is as the characteristic polynomial of multiplication by ξ with respect to the integral basis.

[16]Sanity check: taking $(x_1, x_2, x_3) = (1, 0, 0)$ in the equation above gives $x^3 - 3x^2 + 3x - 1 = (x-1)^3$. Using $(0, 1, 0)$ gives $x^3 - x^2 - 2x - 8$, the minimal polynomial of α, and taking $(0, 0, 1)$ we get $x^3 + x^2 + 2x - 8$, the minimal polynomial of $\beta = 4/\alpha$.

[17]In our case it is

$$\mathfrak{D} = -2012x_2^6 + 2012x_3x_2^5 + 1509x_3^2x_2^4 + 3018x_3^3x_2^3 - 2515x_3^4x_2^2 - 2012x_3^5x_2 - 2012x_3^6,$$

which factors as

$$\mathfrak{D} = -503(2x_2^3 - x_3x_2^2 - x_3^2x_2 - 2x_3^3)^2.$$

as expected.

[18]Kronecker claims this is true, but does not give a full proof. In the number field case, it was proved by Hensel in 1894.

where the polynomials in $n+1$ variables $f_i(x)$ are irreducible modulo p. He then states[19] that p itself, considered as a divisor, is equal to the product of the divisors defined by $p + f_i(\xi)$.[20] Kronecker's actual statement involves divisors of different levels, a notion that he introduces to deal with the higher-dimensional situation. (See [**116**, pp. 459–469] for a discussion.) As indicated in the first words of the passage that follows, and as Hensel's proof eventually showed, this gives the correct factorization in the number field case. So the upshot is: the factorization of a rational prime p in a genus domain is determined by the factorization of the fundamental equation modulo p. This is Kronecker's version of "Dedekind's Theorem."

5.4. Translation

[We translate from Kronecker's *Grundzüge*, section 25, beginning with "Tritt an Stelle eines..." (This is page 382 in [**98**, v. II], page 117 in in the Kummer *Festschrift* and in *Crelle*, vol. 92. We typically give the page numbers from [**98**] and indicate the corresponding pages in *Crelle* in footnotes.) This passage is one long paragraph, but we have broken it up to make the English text more readable. Early on Kronecker refers to equations (A'), (A^0), (X^0); the references can safely be ignored.]

If for the general domain of rationality[21] $[\mathfrak{R}', \mathfrak{R}'', \mathfrak{R}''', \ldots]$ we substitute the ordinary absolute integers[22], then P is a prime number and the quantity $\mathfrak{R}$ introduced above is the only variable.[23] There are no forms higher than the second level[24] and the decomposition[25] (A') of the second-level form[26] $P + \mathfrak{v}\mathfrak{F}(\mathfrak{R})$ into its irreducible factors is therefore valid from the point of view of absolute equivalence (§22, X^0). The key thing for that factorization is the factorization of the congruence $\mathfrak{F}(\mathfrak{R}) \equiv 0 \pmod{P}$. Using that, finding the factorization (A^0) of a prime number

[19]Kronecker's statement is more complicated, introducing new variables and taking into account divisors of different levels. Furthermore, the proof Kronecker gives does not seem to be general. We have opted, then, for giving the key result in the number field case, as stated and proved by Hensel in [**79**]. One can also find the same statement and a similar proof in [**87**, ch. 4].

[20]In our example,

$$F(x) \equiv (x + x_1 + x_2)(x + x_1)(x + x_1 + x_3) \pmod{2}.$$

This implies 2 factors as the product of three divisors, corresponding to

$$2 + 2x_1 + (\alpha + 1)x_2 + \beta x_3,$$

$$2 + 2x_1 + \alpha x_2 + \beta x_3,$$

and

$$2 + 2x_1 + \alpha x_2 + (\beta + 1)x_3,$$

which correspond to Dedekind's $\mathfrak{a}$, $\mathfrak{c}$, and $\mathfrak{b}$, respectively. Notice that these "forms" are not homogeneous.

[21]Here the $\mathfrak{R}$s are the indeterminates generating the base field, the domain of rationality. The notation with square brackets actually means the ring of integers in the base field.

[22]So Kronecker is going to focus, in this section, on $\mathbb{Q}$ as the base field and $\mathbb{Z}$ as the base ring.

[23]This is the x we used in our summary.

[24]"Stufe." Levels were introduced to deal with the more complicated situation, so they are irrelevant over $\mathbb{Z}$.

[25]This is the expression of the divisor corresponding to p as a product of prime divisors obtained by factoring the fundamental equation modulo p.

[26]Since P here is a rational prime, this is just p considered as a divisor in our genus domain, the equivalent of Dedekind's "principal ideal generated by p."

P into its algebraic prime divisors or (following Kummer's language) into its ideal prime factors is both easy and natural.

As mentioned in § 19 (p. 325[27]), at the beginning of my work on algebraic numbers, I attempted to use the theory of higher congruences.[28] If one avoids the factors of the discriminant,[29] this approach succeeds in the most basic and easiest way. In that case, extending of Kummer's definition of ideal numbers to integral functions of the root of an integral equation $\varphi(x) = 0$ is just a matter of checking linear congruences that will tell us whether the resultant of two integral polynomials $\psi(x)$ and $\varphi(x)$ is divisible by a prime number p. If so, $\varphi(x)$ and $\psi(x)$ have a common divisor when considered modulo p. Then those congruence conditions are simply that one of the irreducible factors of $\varphi(x)$, in the sense of congruence modulo p, must be a divisor of $\psi(x)$. This is essentially a reformulation of elementary results developed by Schönemann* in the paper which we quoted above in § 21 (p. 338).[30]

Things are quite different, however, if the prime factors of the discriminant are taken into consideration. An attempt to deal with these in a theory that is based on that limited representation of complex numbers (as polynomial functions of a single algebraic number) is made in the work of Zolotarev published in the latest volume of Resal's journal.[31] This attempt, however, is in my opinion misguided. In fact, Zolotarev cites in the introduction to his paper a publication of Dedekind[32] from the year 1871, which shows clearly and sharply the necessity to abandon this limited view of complex number theory. To attempt to retain it as a starting point nevertheless appears hopeless and contrary to the nature of things. One cannot work with any special equation of an algebraic number, but rather only with the fundamental equation[33] whose formation is based, on the one hand, on the construction of a fundamental system and, on the other hand, on the association of "forms." With this approach the connection to the theory of higher congruences is successful.

[27]The page number refers to the text in **[98]**; it is p. 69 in *Crelle* and the Kummer Festschrift.

[28]As noted, both Dedekind and Kronecker claim to have started their work by trying to generalize Kummer's congruence-based approach. The text in § 19 includes "Die Schwierigkeit, welche die ausserwesentlichen Primfactoren der Discriminante – die ich wegen dieses unregelmässigen Verhaltens damals als 'irregulär' bezeichnete – bei der Zerlegung der Congruenz $F(z) \equiv 0$ darboten, suchte ich Anfangs dadurch zu beseitigen, dass ich andere Gleichungen derselben Gattung zu Grunde legte." ("I tried at first to eliminate the difficulty presented by the inessential prime factors of the discriminant – which I called 'irregular' because of this anomalous behavior – in the analysis of the congruence $F(z) \equiv 0$ by using other equations defining the same genus.") Kronecker, like Dedekind, initially imagined that one might avoid CIDDs by choosing different generators.

[29]This means the discriminant of the minimal polynomial of α, where α is the algebraic integer defining our field.

[30]This is **[125]**. Kronecker's point is that factorization modulo p is computationally feasible and unique, so that the factorization of p is easily found when p does not divide the discriminant of the defining polynomial. Page 338 in **[98]** is page 80 in *Crelle* and the Festschrift.

[31]Resal was then the editor of the *Journal des Mathématiques Pures et Apliquées*, often known as "Liouville's journal." The reference is to **[144]**.

[32]This must the *Anzeige*; see Chapter 3.

[33]"Fundamentalgleichung." See the discussion above.

**Journal f. Mathematik* Bd. 31, S. 269

In the case considered here of the realm of absolute rationality,[34] the discriminant form is equivalent[35] to an ordinary integer, which we can denote by D. The ratio $\mathfrak{D}/D$ is then a primitive form with indeterminate u,[36] except[37] that, in case the genus domain $\mathfrak{G}$ has conjugates, it can be multiplied by powers of degree less than $n-1$ of the prime divisors of D.

In that case,[38] of course, the discriminant of each specific numerical equation of the realm $\mathfrak{G}$ can have superfluous factors, but even when this is not the case and the quotient $\mathfrak{D}/D$ is a primitive form without numerical factor, it can happen that all numerical values of the indeterminate u will yield values that share a common factor. For example, this occurs when the primitive form is an integral homogeneous function of terms $u^p - u$ with p a prime (see § 8 at the end, p. 273).[39]

One such case is the genus domain given by Mr. Dedekind,[40] defined by the cubic equation $\alpha^3-\alpha^2-2\alpha-8=0$.* In this case, if u is an indeterminate, $\alpha+4u/\alpha$ is a root of the fundamental equation, and its discriminant is D times the square of the primitive form[41] $2u^3+u^2+u-2$, whose values are divisible by 2 for all integer values of u.

I myself found a similar example in the thirteenth roots of unity[42] when, in 1858, I was doing my first work on this theory. These examples suffice to show that the association[43] of forms corresponds to the nature of things, not only in the general theory of algebraic quantities, but also in the more specific theory of algebraic numbers. But it should be mentioned here that for this elementary question of the factorization of complex numbers, completely analogously to the harder question of the representation of prime divisors, a second type of association[44] can be used, namely with higher algebraic numbers. In fact, it is easily seen that the unknowns

[34]In other words, working over $\mathbb{Q}$.

[35]This is the discriminant of the fundamental equation, hence a "form," i.e., a polynomial. But any polynomial in $\mathbb{Z}[x]$ is equivalent, in Kronecker's sense, to its content.

[36]Perhaps Kronecker means "indeterminates" here; we do not see why there would be only one variable. But equivalence does sometimes allow reducing the number of variables; see below.

[37]We think what is going on here is this. Most of Kronecker's theory in this section is developed under the assumption that the genus domain is normal, so that all the conjugates or its elements are available. If not, then the "genus domain has conjugates," i.e., conjugating all the elements gives another genus domain. Then Kronecker has to worry about the difference between the discriminant of the field and that of its normal closure, which is what provides these extra factors.

[38]I.e., when the field is not normal.

[39]Page 25 in *Crelle* and the Festschrift.

[40]The reference is to the paper we translated in Chapter 4 **[23]**, but this is the same polynomial Dedekind found in 1871 **[20]** and the one we worked with in the example above. If Kronecker only read the 1878 paper he likely missed the fact that Dedekind had actually constructed infinitely many examples, which is clear in the 1871 *Anzeige*.

[41]Kronecker gives no details of the computation here, nor does he explain how he has reduced from three variables to one. As shown above, $\mathfrak{D} = -503(2x_2^3 - x_3x_2^2 - x_3^2x_2 - 2x_3^3)^2$. To get Kronecker's version, divide by $-x_2^3$ and let $u = x_3/x_2$.

[42]Judging from what Hensel did later, this means that the example is a subfield of $\mathbb{Q}(\zeta_{13})$, in fact the unique quartic subfield. See Section 2.5.

[43]In this context, "association" seems to translate to base change, from $\mathbb{Z}$ to $\mathbb{Z}[u_1, u_2, \dots]$ or to Penchèvre's localization.

[44]Now Kronecker wants to base change from $\mathbb{Z}$ to the ring of integers of some auxiliary field.

***R. Dedekind*, "Ueber den Zusammenhang zwischen der Theorie der Ideale und der Theorie der höheren Congruenzen." Abhandlungen der Köngl. Gesellschaft der Wissensch. zu Göttingern. Bd. 23, S. 30.

of that primitive form which is a root of the fundamental equation can always be given[45] algebraic number values that will serve the stated purpose.[46] For the example above, if we take u to be a third root of unity we get a primitive form whose absolute value is equal to one.[47]

[There is a break at this point (p. 384 in [**98**], p. 119 in *Crelle.*), after which Kronecker returns to the general case.]

5.5. Kronecker and CIDDs

It is noteworthy that Kronecker, like Dedekind, points to the existence of common inessential discriminant divisors as sufficient evidence that one cannot be satisfied with basing the theory of algebraic numbers on higher congruences. In § 19, he tells us that he attempted to get around inessential divisors by "using another equation," an attempt that failed.

Dedekind's 1871 example, he says,

> ...shows clearly and sharply the necessity to abandon this limited view of complex number theory. To attempt to retain it as a starting point nevertheless appears hopeless and contrary to the nature of things. One cannot work with any special equation of an algebraic number, but rather only with the fundamental equation whose formation is based, on the one hand, on the construction of a fundamental system and, on the other hand, on the association of "forms." With this approach the connection to the theory of higher congruences is successful.

This declaration is motivated by Zolotarev's theory, of course. Notice that Kronecker does not claim Zolotarev's theory is incorrect. Rather, he argues that to insist on an approach via "higher congruences" is a mistake even if we can work around the problem: it is against "the nature of things."

Kronecker refers to "the work of Zolotarev published in the latest volume of Resal's journal." This is the paper in French published in 1880 [**144**], so there was no language barrier. Did Kronecker read it? Since the paper was actually published in three parts, it is possible that he only read part of it. In any case, it is clear that Kronecker is reacting to Zolotarev's use of higher congruences, but he seems unaware of the "local" methods introduced later in the paper.

Kronecker claims to have realized early on that CIDDs exist, when he first started working on algebraic numbers, even before Kummer's remark in 1859. He does not specify much about his example beyond that it is "in the thirteenth roots

[45]This is, we think, what Ore calls "Kronecker's conjecture". We know the discriminant of the fundamental equation is equal to $D\Delta^2$, where $D \in \mathbb{Z}$ and Δ is a polynomial in several variables with integer coefficients. In the case where common inessential divisors exist, when we give integer values to the variables the value of Δ will always be divisible by some prime. Kronecker here suggests that if instead we take values in some more general ring of integers, those common factors will disappear. Hensel proved this in 1894; see the next chapter.

[46]Kronecker says "In der That sieht man leicht, dass die Unbestimmten jener primitiven Form, welche eine Wurzel der Fundamentalgleichung darstellt, stets als ganze algebraische Zahlen dem angegebenen Zwecke gemäss bestimmt werden können." We think the "angegebenen Zwecke" is to eliminate the extra factor(s).

[47]Indeed, is $u = \zeta_3$ then $2u^3 + u^2 + u - 2 = -1$.

of unity." We suggested a reconstruction of this example above; see Section 2.5. Kronecker left it to Hensel to work out the details and generalize his example.

The consensus of the time seems to have been that Kronecker's proofs in Section 25 we not complete. In the number field case, the first proofs were given by Hensel in 1894, after Kronecker's death. (It is possible that Hensel expected to find the proofs among Kronecker's unpublished papers. When he did not, he wrote up his own version.) Hensel followed that paper with a second one giving his solution to the problem of common inessential discriminant divisors. We translate and analyze that paper in the next chapter.

CHAPTER 6

Hensel's 1894 Paper on Common Inessential Discriminant Divisors

Kurt Hensel's mathematics often appears old-fashioned to modern eyes. Born in 1861, Hensel[1] actually belonged to the same generation as David Hilbert, who was born in 1862. As a student of Leopold Kronecker, however, Hensel remained close to his teacher's mathematical style, though perhaps more because of personal loyalty than for philosophical reasons.

Hensel began working under Kronecker in the early 1880s.[2] Hensel dedicated his thesis to the problem of common inessential discriminant divisors, apparently without first reading Dedekind's 1878 paper — or at least without having read it carefully. It appears that he began from Kronecker's vague memory of having run into CIDDs "in the thirteenth roots of unity," worked out the details of that example, and then went on to attempt generalizations.

Hensel's thesis was entitled *Arithmetische Untersuchungen über Discriminanten und ihre ausserwesentlichen Teiler* (Arithmetical investigations of discriminants and their inessential divisors). A thesis booklet **[74]**, published in 1882, has two parts. In a twelve-page introduction, Hensel summarizes what he knows about CIDDs, from Dedekind's example to various theorems he claims to have found. The remaining eighteen pages, however, present proofs of only a small portion of the results announced in the introduction. "I hereby conclude this preliminary publication, in which only an overview of the contents of my dissertation submitted to the faculty," Hensel writes near the end, adding later that the "complete publication should follow as soon as possible."

In his book **[10]** on the Berlin mathematics department and its professors, Biermann includes as document 14 the official report on Hensel's thesis and doctoral examination. In it Kronecker claims (incorrectly,[3] as we know from Chapter 3) that before Hensel's "unusually extensive" (außerordentlich umfangreiche) dissertation, only isolated examples were known of fields in which common inessential discriminant divisors exist. He highlights that Hensel's results suffice to show there are infinitely many such fields.

[1]The major source of information about Kurt Hensel's life is Helmut Hasse's memorial article **[71]**. Based on notes provided by Gertrud Hensel and on Hasse's own memories, it gives both a summary of the "outer life" and an overview of Hensel's mathematical work. The first chapter of Birgit Petri's thesis **[117]** complements and completes Hasse's account, using primary sources whenever possible.

[2]Petri **[117]** had access to a mathematical diary from Hensel's early student days (1879–1883), so her account of Hensel's early work is particularly valuable. Our account is mostly based on her work.

[3]As we suggested, it is possible that Kronecker only read Dedekind's 1878 paper (translated in Chapter 4). There Dedekind limits himself to giving one example and does not explain how it was constructed.

Following the completion of his doctorate, Hensel continued to work on topics inspired by Kronecker's *Grundzüge*, which was his major source. Hensel refers to the *Grundzüge* as "Kronecker's *Festschrift*" because Kronecker originally published it to commemorate the 50th anniversary of Kummer's doctorate.

Hensel submitted several, mostly unpublished, papers for his Habilitation in 1886.[4] We know their content only from Kronecker's notes. The Habilitation papers seem to form the basis of several of Hensel's subsequent papers. No *Habilitationsschrift* was published in 1886.

Kronecker died in 1891. It is possible that Hensel had held back on publishing his own work because he expected to find proofs for many of his results in Kronecker's papers. But he did not; see the first paragraph of Section 6.4, §4 below. In particular, it was necessary to give proofs of some of the results Kronecker had claimed in the *Grundzüge*, as discussed in the previous chapter.

Hensel published two important papers in 1894. Petri has argued in [**117**, 2.4] that both of them contain material that was originally submitted for Hensel's Habilitation in 1886, but on which he had continued to work in the following years.

6.1. Hensel's First 1894 Paper

The first paper Hensel published in 1894 had the title "Investigation of the Fundamental Equation of a Genus Modulo a Real Prime and Determination of the Numerical Divisor of its Discriminant."[5] This points directly to Section 25 of the *Grundzüge*. We are considering an algebraic number which is a root of a monic irreducible polynomial of degree n with integer coefficients. The "fundamental equation" of the genus domain it determines is the degree n polynomial $F(x)$ with coefficients in $\mathbb{Z}[u_1, u_2, \ldots, u_n]$ satisfied by the fundamental form (or "generic integer")

$$\xi = u_1\omega_1 + u_2\omega_2 + \cdots + u_n\omega_n,$$

where the ω_i are an integral basis. We have $F(\xi) = 0$ and the polynomial $F(x)$ is irreducible. Recall that Kronecker had said, in Section 25 of the *Grundzüge*, that the factorization of a rational prime p in the genus domain can be determined by factoring the polynomial $F(x)$ modulo p.

There are four main theorems in Hensel's [**79**].

(a) For any rational prime p, the congruence modulo p of lowest degree satisfied by ξ is $F(\xi) \equiv 0 \pmod{p}$.

(b) If $D(u_1, u_1, \ldots, u_n)$ is the discriminant of the fundamental equation and d_K is the discriminant of the genus domain, then

$$D(u_1, u_1, \ldots, u_n) = d_K \Delta(u_1, u_2, \ldots, u_n)^2,$$

where $\Delta(u_1, u_2, \ldots, u_n)$ is a polynomial with integer coefficients that have no common factor.

(c) Factoring the fundamental equation $F(x)$ modulo p gives the factorization of p in the genus domain.

(d) A prime p is divisible by the square of a prime divisor of the genus domain if and only if p is a divisor of the genus discriminant d_K.

---

[4] See Petri's reconstruction, [**117**, 2.3].

[5] "Untersuchung der Fundamentalgleichung einer Gattung für eine reelle Primzahl als Modul und Bestimmung der Teiler ihrer Discriminante," see [**79**].

The second theorem was stated by Kronecker in Section 25 of the *Grundzüge*. In modern terms, it says that the field discriminant is the content of the discriminant of the fundamental equation. In Kroneckerian terms, it says that Δ and d_K are equivalent as divisors. The third theorem was also stated by Kronecker, as we saw in the previous chapter. The fourth theorem was stated and proved by Dedekind (statement in the *Anzeige*; see Chapter 3, proof in **[24]**).

Petri gives a detailed summary of the content of **[79]** in her thesis; see **[117**, 2.4.2**]**. She notes, in particular, a comment by Hensel early in the paper:

> How one wants to define these prime factors, whether in closed form, as linear forms with indefinite coefficients as in Kronecker or, as in Dedekind, as the totality of infinitely many definitely characterized algebraic integers, is completely irrelevant for the present investigation. What matters is only that you can see whether an algebraic number is divisible by a prime divisor P_i or not, and this question can easily be decided in both viewpoints.[6]

This, as Petri notes, is a significant claim: the foundation does not matter as long as we obtain a reasonable divisor theory. (It seems likely that Kronecker would not have agreed.) Having made this concession, Hensel nevertheless uses Kronecker's methods throughout. Then, towards the end of his paper, he compares them[7] to Dedekind's approach, citing especially **[24]**, the 1882 paper in which Dedekind shows that the prime divisors of the field discriminant are exactly the ramified primes. Hensel argues that his proofs are much easier than Dedekind's, showing that Kronecker's methods are preferable.

6.2. Hensel's Second 1894 Paper: Criteria for CIDDs

Hensel's second paper from 1894, entitled "Arithmetical Investigations of the Common Inessential Discriminant Divisors of a Genus,"[8] is the text we translate and analyze in this chapter. The goal is to complete the theory of common inessential discriminant divisors and, in particular, to prove several of the results claimed in the introduction to Hensel's thesis. We give a detailed outline below.

As usual, we have chosen to focus on getting the mathematical content right, preserving Hensel's language, notations, and general point of view. We have not tried (and would not have succeeded) to preserve every nuance of meaning or to reproduce Hensel's grammar precisely.

Our translation is based on the original publication in the *Journal für die Reine und Angewandte Mathematik* (**113** (1894), 128–160). Hensel's own footnotes (which are few) are marked by asterisks, while our annotations are given in numbered footnotes.

Hensel indicates theorems by using indented text; we have used a modern simulacrum of the same device. Hensel does not signpost the beginning or end of

[6]Wie man diese Primfactoren definiren will, ob in geschlossener Form, als Linearformen mit unbestimmten Coefficienten, wie bei *Kronecker* oder, wie bei *Dedekind*, als die Gesammtheit von unendlich vielen bestimmt charakterisirten ganzen algebraischen Zahlen, ist für die vorliegende Untersuchung völlig gleichgültig. Es kommt hier nur darauf an, dass man erkennen kann, ob eine algebraische Zahl durch einen Primtheiler P_i theilbar ist, oder nicht, und diese Frage kann bei beiden Auffassungsarten leicht entschieden werden. **[79**, p. 63**]**

[7]See **[79**, pp. 81ff**]**; see **[117**, p. 77**]** for Petri's discussion.

[8]Arithmetische Untersuchungen über die gemeinsamen ausserwesentlichen Discriminantentheiler einer Gattung.

a proof; we have often added such signposts in the footnotes. We have retained Hensel's notation as much as possible.

Every once in a while words have been inserted in square brackets when we felt it would clarify the meaning. Hensel often uses $\gtrless$ to indicate inequality or incongruence; we have silently substituted $\neq$ or $\not\equiv$. We have also rendered "ganze Functionen" and "ganze ganzzahlige Functionen" as "polynomials" and "integral polynomials" or "polynomials with integer coefficients", respectively, without further comment. Hensel often says "order" when we would say "degree"; he sometimes also uses "dimension" in a similar sense. We have mostly translated "degree" when it was unambiguous what was meant; see the footnotes.

Hensel often uses "genus" (Gattung) where a modern mathematician might want something more specific: "genus domain" when he means a field, "species" (Art) or "principal species" when he means the ring of integers, etc. We have often opted for "genus" or "domain," adding a footnote when it seemed advisable.

6.3. Outline of the Paper

Hensel's paper contains five sections which he labels §1 to §5. The results in each section are as follows.

§1. The main theorem here is that a prime p is a common inessential discriminant divisor in a field K if and only if there are not enough irreducible polynomials modulo p to match the factorization of p in K. This result is, of course, also in Dedekind's 1878 paper; see Chapter 4. Hensel cites this paper, but it remains unclear how carefully he had read it. Certainly several of the arguments are identical in substance, but Hensel states and proves the main theorems in terms of congruences mod p, and only later translates them into Dedekind's language of indices. All of these theorems had already been stated in the introduction to Hensel's thesis booklet.

§2. Using the criterion Hensel just found seems to require knowing the factorization of p, but in fact all we need is a way to determine how many prime divisors of p in K have a given degree. Gathering this information for each degree allows us to compare the number of divisors with the number of irreducible polynomials of that degree modulo p. In this section Hensel shows that one can determine these numbers without knowing the full factorization of p. Petri argues[9] that the material in this paragraph was not part of the Habilitation materials, hence was new in 1894.

Section two is remarkable in several ways. Together with Dedekind's criterion from section one, it completely solves the problem of determining whether p is a common inessential discriminant divisor. The method involves constructing a family of special divisors (in Dedekind's language, ideals)[10] using an integral basis. These divisors are studied only "above p," i.e., Hensel ignores (mostly silently) all the prime divisors in his field except those that divide the prime p. (For example, the divisors Hensel constructs depend on the choice of integral basis, but their factorization above p does not.) This "semi-local" approach had been introduced by Hensel in **[75, 76]** in order to study congruences modulo a prime divisor of p.

---

[9]See **[117**, p. 84].

[10]The authors have been unable to find in the literature further discussion of the divisors (or ideals) that Hensel constructs.

It is developed further here, however silently. This would eventually connect nicely to Hensel's p-adic methods, which were still in the future.

§3. The focus in this section changes to the "discriminant of the fundamental equation with its numerical factor removed," i.e., the index form $\Delta(u_1, u_2, \ldots, u_n)$ (with respect to a fixed[11] integral basis). As we know, p is a common inessential discriminant divisor exactly when substituting any n-tuple of integers for the variables in the index form Δ results in a number divisible by p. Hensel derives a general criterion to recognize when a polynomial with integer coefficients has this property. Kronecker had given a sufficient condition for this to happen, which Hensel proves is also necessary. He also explains how to test the condition in any specific case, showing that if we know the index form[12] it is computationally easy to decide whether it has this property. (All of these theorems had also been announced in the introduction to Hensel's thesis booklet.) As an application, Hensel gives a new proof of a result from his thesis characterizing when $p = 2$ is a CIDD in certain cubic fields.

§4. Kronecker had observed (see the previous chapter) an interesting property of the index form in Dedekind's cubic field example. While every value obtained by taking integer values for the u_i in $\Delta(u_1, u_2, u_3)$ was divisible by 2, that was no longer the case if we allowed the u_i to take values that are integers from the cyclotomic field $L = \mathbb{Q}(\zeta_3)$. In this section Hensel shows that for any polynomial with integer coefficients we can find an auxiliary field L with this property.

Kronecker was very interested in "auxiliary fields" of several kinds,[13] sometimes anticipating class field theory. In this case it seems unclear whether something deep is going on in the "association" of L to the given index form. The natural thing, in modern terms, would be to understand this in terms of the field extensions $K/\mathbb{Q}$ and KL/L. Neither Kronecker nor Hensel, however, consider whether the index form attached to K over $\mathbb{Q}$ is in any way related to the extension KL/L. They simply focus on the form and its values. Hensel will nevertheless use phrases such as "the prime p will be a common inessential divisor in the domain of rationality L" when substituting integers from L into the index form always gives results that are divisible by p.

§5. Given the result in §4, it is natural to ask which field we need to use. Hensel shows that one can always choose a subfield of a cyclotomic field of prime order. (He describes this as finding "the supplementary genus of the greatest algebraic simplicity.") The key fact, in modern terms, is that only the residue field of $\mathcal{O}_L$ at a prime above p is relevant.

While the results in sections one and three are well known, the material in sections two, four, and five has mostly not been studied. We think that the ideas in section two, in particular, deserve more attention. The ideals F_κ might well be of independent interest. The fact that Hensel's arguments anticipate local methods shows that Hensel already understood that he could study the situation "at p," even though he only says something along those lines once (see footnote 89).

[11] A choice of integral basis is everywhere in Kronecker's approach, another aspect Dedekind would have disliked.

[12] Potentially a big "if."

[13] See his remarks about "association" towards the end of Section 6.4 (on page 86).

6.4. Translation

Arithmetical Investigations of the Common Inessential Discriminant Divisors of a Genus

(by Mr. K. Hensel)

---

§1

In a recently published work (this Journal, vol. 111[14], pp. 61–83) I considered the congruence of least degree modulo a prime p satisfied by the fundamental form[15]

$$(1.)\qquad w_0 = u_1\xi_1 + \cdots + u_n\xi_n$$

of a given genus domain[16] of the n-th order.[17] The main result, which will serve as the basis for this work, says that, for any prime number, w_0 does not satisfy a congruence of degree smaller than the degree n of the genus.[18]

The congruence of lowest degree which is satisfied by w_0 modulo p is made up in a simple manner from the congruences satisfied by w_0 modulo each of the prime divisors of p. Let P be one of these factors in the domain[19] $(\mathfrak{G})$, and let κ be its degree.[20] Then w_0, with indeterminates $u_1, \ldots, u_n$, satisfies (modulo P) the congruence of degree κ

$$(2.)\qquad \mathfrak{F}(w) = w^\kappa + U^{(1)}(u_1 \ldots u_n)w^{\kappa-1} + \cdots + U^{(\kappa)}(u_1 \ldots u_n) \equiv 0 \pmod{P},$$

whose left side[21] is irreducible modulo the prime p, while the coefficients are integral polynomials in $u_1 \ldots u_n$.

Let then

$$(3.)\qquad p = P_1^{\delta_1} P_2^{\delta_2} \ldots P_h^{\delta_h}$$

---

[14]This is a typo, as noted on page 160 of this issue; it should be Volume 113. The paper is **[79]**.

[15]This alerts the reader that the u_i in this equation are supposed to be indeterminates. Hensel, following Kronecker, uses "form" to mean a polynomial in several variables. The ξ_i in this expression are what we would call an integral basis. The entire argument assumes an integral basis has been chosen. As discussed above, w_0 is a root of the fundamental equation $\mathfrak{F}(w) = 0$ of the genus domain. Hensel often refers to it as the "equation of smallest degree satisfied by w_0." Notice that w_0 is the fundamental form and w is the variable in the polynomial it is a root of. Hensel will follow this notational pattern throughout.

[16]"Gattungsbereich"

[17]We would say "degree" instead of "order."

[18]In modern terms, the element w_0, considered modulo p, is integral of degree n over $\mathbb{F}_p[u_1, u_2, \ldots, u_n]$.

[19]"Bereich." Hensel follows Kronecker in using $(\mathfrak{G})$ as the notation, which means that $\mathfrak{G}$ is the element generating the genus domain over the rationals.

[20]"Ordnungzahl." We would call it the residual degree of P. Hensel defined this number in **[79]** and showed that the number of distinct congruence classes modulo P in the domain is p^κ.

[21]Hensel thinks in terms of equations and congruences, not polynomials. The "left hand side" of the congruence $\mathfrak{F}(w) \equiv 0$ is the minimal polynomial for w_0 (mod P). It is actually the middle term in the double congruence given here.

be the decomposition of p into its prime factors in the domain $(\mathfrak{G})$ and let

$$\mathfrak{F}_1(w),\ \mathfrak{F}_2(w),\ \ldots,\ \mathfrak{F}_k(w) \tag{4.}$$

be the functions[22] of lowest degree having the fundamental form w_0 as a root modulo the h corresponding distinct prime divisors

$$P_1,\ P_2,\ \ldots,\ P_h. \tag{5.}$$

The polynomials have (as polynomials in w) degrees

$$\kappa_1,\ \kappa_2,\ \ldots,\ \kappa_h,$$

which are equal to the degrees of the prime factors of p.[23] Then in the cited paper[24] (p. 75) it is shown that the congruence of lowest degree satisfied by w_0 modulo the prime number p will be the following:

$$\mathfrak{F}_1^{\delta_1}(w)\mathfrak{F}_2^{\delta_2}(w)\ldots\mathfrak{F}_h^{\delta_h}(w) \equiv 0 \pmod{p}, \tag{6.}$$

and its degree in w is

$$\kappa_1\delta_1 + \kappa_2\delta_2 + \cdots + \kappa_h\delta_h = n. \tag{6^{a}.}$$

If we plug in w_0 for w, each of the $\mathfrak{F}_i(w_0)$ is divisible by the divisor P_i, so the whole product is divisible by $P_1^{\delta_1}\ldots P_h^{\delta_h}$ and so divisible by p.

Let us now consider, instead of the fundamental form w_0, an algebraic integer of the genus domain $(\mathfrak{G})$

$$\xi_0 = a_1\xi_1 + a_2\xi_2 + \cdots + a_n\xi_n, \tag{7.}$$

which therefore is simply obtained[25] from w_0 by giving the indeterminates $(u_1,\ldots,u_n)$ the integer values $(a_1,\ldots,a_n)$. We would like to investigate which congruence with integer coefficients is satisfied by ξ_0 modulo p.

Obviously ξ_0 satisfies the congruence of n-th degree given by (6.), since we are just replacing $(u_1,\ldots,u_n)$ by $(a_1,\ldots,a_n)$. In general,[26] if we have polynomials

$$\mathfrak{F}_1(w),\ \mathfrak{F}_2(w),\ \ldots,\ \mathfrak{F}_h(w)$$

of degrees $\kappa_1,\kappa_2,\ldots,\kappa_h$, we write the corresponding[27] functions

$$\mathfrak{F}_1(\xi),\ \mathfrak{F}_2(\xi),\ \ldots,\ \mathfrak{F}_h(\xi)$$

(where the variable w is now replaced by ξ to indicate the difference). Then ξ_0 is a root of the congruence of degree n with integer[28] coefficients

$$\mathfrak{F}_1^{\delta_1}(\xi)\ldots\mathfrak{F}_h^{\delta_h}(\xi) \equiv 0 \pmod{p}. \tag{8.}$$

---

[22] In general, Hensel means "polynomial" when he says "function."

[23] So Hensel knows that the factorization of p is determined by the factorization mod p of the fundamental equation, which is the polynomial with coefficients in $\mathbb{Z}[u_1,u_2,\ldots,u_n]$ having w_0 as a root.

[24] This is **[79]**.

[25] So ξ_0 is a particular algebraic integer, obtained as a linear combination of the integral basis $\xi_1,\xi_2,\ldots,\xi_n$.

[26] "In general" here seems to mean that this notation will always be used.

[27] Hensel is introducing notation: these are the same functions as before, except that he has replaced the indeterminates u_i by the integers a_i. The different variable ξ indicates that this has been done. Notice that the functions $\mathfrak{F}_i(\xi)$ therefore depend on our choice of ξ_0, i.e., depend on the choice of the integers a_i. The main point of the notation is precisely not to have to show the dependence on the u_i or the a_i.

[28] Whereas in (6.) the coefficients were polynomials in n variables.

We know this to be the case because the individual numbers $\mathfrak{F}_i(\xi_0)$ are obtained from the corresponding forms $\mathfrak{F}_i(w_0)$ by replacing the unknowns $(u_1, \ldots, u_n)$ by the integers $(a_1, \ldots, a_n)$. Therefore each of these are divisible a fortiori by the each of the prime factors as before.

The fundamental form w_0, as proved in the aforementioned work, does not satisfy any congruence whose degree is smaller than that of (6.). But the congruence (8.) need not be the congruence of lowest degree[29] satisfied by ξ_0; for that to be true we would need to choose ξ_0 appropriately. We must then investigate the following question:

Under what conditions will the algebraic integer ξ_0 satisfy no congruences modulo p of degree less than n?

It is very easy to give a system of necessary conditions; we will later prove that they are also sufficient. First,[30] the h integer functions of ξ in (8.), $\mathfrak{F}_1(\xi), \ldots \mathfrak{F}_h(\xi)$ must be irreducible modulo p. In fact,[31] if for example

$$\mathfrak{F}_1(\xi) \equiv F_1(\xi).G_1(\xi) \pmod{p},$$

where F_1 and G_1 are functions of ξ of degree lower than κ_1, then the same congruence also holds modulo the prime divisor P_1 of p. Substituting ξ by ξ_0 and observing that $\mathfrak{F}_1(\xi_0)$ is divisible by P_0, we see the congruence

$$F_1(\xi_0).G_1(\xi_0) \equiv 0 \pmod{P_1},$$

from which it follows that one of these factors, say $F_1(\xi_0)$, is divisible by the prime divisor P_1. Then ξ_0 clearly satisfies the congruence modulo p

$$(8^{\mathrm{a}}.) \qquad F_1^{\delta_1}(\xi)\mathfrak{F}_2^{\delta_2}(\xi) \ldots \mathfrak{F}_h^{\delta_h}(\xi) \equiv 0 \pmod{p},$$

because the first factor on the left side is divisible by $P_1^{\delta_1}$, while the others are divisible by $P_2^{\delta_2}, \ldots, P_h^{\delta_h}$. But the degree of (8^{a}.) is smaller than n, since it is smaller[32] than the degree of (8.).

Secondly, the h irreducible polynomials $\mathfrak{F}_i(\xi)$ must be distinct[33] modulo p. If for example[34] $\mathfrak{F}_1(\xi)$ and $\mathfrak{F}_2(\xi)$ were congruent for this modulus, then a fortiori

$$\mathfrak{F}_1(\xi) \equiv \mathfrak{F}_2(\xi) \pmod{P_1P_2},$$

since P_1P_2 is a divisor of p. Substituting again ξ_0 for ξ and noticing that $\mathfrak{F}_1(\xi_0)$ is divisible by P_1 and $\mathfrak{F}_2(\xi_0)$ is divisible by P_2, it follows from the above congruence that $\mathfrak{F}_1(\xi_0)$ and $\mathfrak{F}_2(\xi_0)$ are both divisible by the product (P_1P_2). Now if we take any two exponents δ_1, δ_2 with $\delta_1 \geq \delta_2$, the power $\mathfrak{F}_1^{\delta_1}(\xi)$ will, when $\xi = \xi_0$, be divisible by the product $(P_1P_2)^{\delta_1}$, so a fortiori by $P_1^{\delta_1}P_2^{\delta_2}$. Thus ξ_0 satisfies the congruence

$$(8^{\mathrm{b}}.) \qquad \mathfrak{F}_1^{\delta_1}(\xi)\mathfrak{F}_3^{\delta_3}(\xi) \ldots \mathfrak{F}_h^{\delta_h}(\xi) \equiv 0 \pmod{p},$$

---

[29] For example, if we choose all of the a_i divisible by p, the congruence of lowest degree will be the degree one polynomial ξ.

[30] Hensel doesn't do Lemmas, but this is one. Formally: if ξ_0 does not satisfy any congruence of degree less than n, then the polynomials appearing in (8.) must be irreducible modulo p. Notice that these are the polynomials *after* substituting the u_i by the a_i, so they depend on the choice of ξ_0.

[31] Here begins the proof of the Lemma.

[32] The contradiction ends the proof of the Lemma.

[33] Lemma 2, still under the running assumption that ξ_0 satisfies no congruence of degree lower than n.

[34] Here begins the proof.

whose degree is smaller than that of (8.), so smaller[35] than n. So we have the following result:

(A.) If the number ξ_0 does not satisfy any congruence of degree less than n modulo p, then the h polynomials $\mathfrak{F}_1(\xi), \dots, \mathfrak{F}_h(\xi)$ in (8.) are all distinct and irreducible modulo p.

This theorem was given by Mr. *Dedekind* in his major work[36] "Ueber den Zusammenhang zwischen der Theorie der Ideale und der Theorie der höheren Congruenzen" (*Abh. der Gött. Gesellschaft* Volume 23), although in slightly different form.[37] It is noteworthy that he demonstrated that for it to be possible to find such h functions $\mathfrak{F}_1(\xi), \dots, \mathfrak{F}_h(\xi)$, it is necessary and sufficient that a number ξ_0 exists for which p is not an inessential divisor of the discriminant.[38] This result[39] enabled him to find a specific genus of degree three for which the prime number 2 is a common inessential divisor of all equation discriminants. This alone shows that that the theory of genus domains[40] cannot be founded upon higher congruences. This[41] can be done, however, if, as in this and the previous work, we work with with the linear *form* $w_0 = u_1\xi_1 + \cdots + u_n\xi_n$ with indeterminates $u_1, \dots, u_n$, rather than a specific *number* ξ_0 of the domain. This is because in the previous work[42] it was indeed established that in the discriminant of the polynomial having w_0 as a root no prime p is contained as other than an essential divisor.[43] The results on common inessential divisors of the discriminant that follow in this paper have not, to my knowledge, been given before.[44]

We now want to investigate when condition (A.) can actually hold.[45] If the h polynomials $\mathfrak{F}_1(\xi), \dots, \mathfrak{F}_h(\xi)$ are irreducible modulo p, two of them, say $\mathfrak{F}_1(\xi)$ and

[35] End of the proof, again by contradiction.

[36] The original is "grossen Arbeit." This is **[23]**, which we translated in Chapter 4. It is not really a full-length memoir, so "grossen" seems unlikely to mean "large." "Great work" seemed excessive, however.

[37] Dedekind does not use the language of satisfying congruences in **[23]**, but the connection can be found implicitly there. Hensel is assuming a connection he has not yet shown between "satisfying a congruence of degree less than n" and "having index divisible by p."

[38] This is Theorem (IV) in **[23]**; see Chapter 4. So at this point Hensel seems to know Dedekind's 1878 criterion for p to be a common inessential discriminant divisor. The logical structure here is this, therefore: if (1) an element ξ_0 does not satisfy a congruence of degree less than n, then (2) we must have h distinct irreducible polynomials modulo p, which is equivalent (apud Dedekind) to (3) p is not a common inessential discriminant divisor in this field.

[39] "This result" must be Dedekind's criterion for p being a CIDD rather than the necessary criterion in Theorem A, which is about satisfying congruences.

[40] "Theorie der Gattungen." The same observation is made by Dedekind implicitly in 1871, explicitly in the 1878 paper (see, for example, the footnote at the end of §3 of his paper translated in Section 4.3, page 62) and by Kronecker (see his prefatory remarks in Section 5.4). This is exactly what Zolotarev tried to do but had to move beyond. Note the consensus here: the several different pioneers of algebraic number theory point to this issue to justify why the "simple and direct" approach must be abandoned.

[41] I.e., basing the whole theory on polynomial congruences.

[42] The reference is to **[79]**, summarized above.

[43] The theorem is that if you compute the discriminant of the fundamental equation you obtain a polynomial in n variables u_i whose content is d_K.

[44] It is not clear which "results that follow" Hensel claimed were new. Did he mean all but Theorem A? If so, he was wrong: all of the results in §1 were in Dedekind's paper, but expressed in terms of the index $(\mathcal{O} : \mathbb{Z}[\theta])$ instead of in terms of "the lowest degree congruence."

[45] This section is about counting how many irreducible polynomials are available for Theorem A. The theorem and the method were already well known; for example, it can be found

$\mathfrak{F}_2(\xi)$, can only be congruent if they have the same degree, i.e., if the degrees κ_1 and κ_2 of the corresponding prime factors P_1 and P_2 are equal. So let us arrange the distinct prime divisors $P_1, P_2, \dots, P_h$ by their degrees $\kappa_1, \kappa_2, \dots \kappa_h$, grouping together those that have equal degrees. Of the h integers, suppose that there are

$$\lambda_1 \text{ equal to } \kappa_1, \quad \lambda_2 \text{ equal to } \kappa_2, \quad \dots \quad \lambda_\gamma \text{ equal to } \kappa_\gamma,$$

where $\lambda_1 + \lambda_2 + \cdots + \lambda_\gamma = h$ and the degrees $\kappa_1, \kappa_2, \dots, \kappa_\gamma$ are all different. For the moment, take κ to be one of these γ degrees[46] $\kappa_1, \dots, \kappa_\gamma$ and let

$$(9.) \qquad P^{(1)}, P^{(2)}, \dots, P^{(\lambda)}$$

be the prime factors of p whose degree is equal to κ. Likewise, let

$$(9^{\mathrm{a}}.) \qquad \mathfrak{F}^{(1)}(\xi), \mathfrak{F}^{(2)}(\xi), \dots, \mathfrak{F}^{(\lambda)}(\xi)$$

be functions of degree κ satisfied by ξ_0 modulo the prime factors $P^{(1)}, \dots P^{(\lambda)}$. We now want to know if it is possible for these λ functions to be irreducible and incongruent modulo p.

Since $\mathfrak{F}^{(1)}(\xi)$ is irreducible [and has ξ_0 as a root] modulo $P^{(1)}$, if ξ_0 satisfies another polynomial congruence[47]

$$\Phi(\xi) \equiv 0 \pmod{P^{(1)}},$$

then $\Phi(\xi)$ must be divisible by $\mathfrak{F}^{(1)}(\xi)$ modulo $P^{(1)}$, because otherwise $\Phi(\xi)$ and $\mathfrak{F}^{(1)}(\xi)$ would have a greatest common divisor modulo $P^{(1)}$, which contradicts by the irreducibility of $\mathfrak{F}^{(1)}(\xi)$. Therefore for the modulus $P^{(1)}$ and therefore[48] for p itself, we get a congruence of the form:

$$\Phi(\xi) \equiv \mathfrak{F}^{(1)}(\xi)\Phi^{(1)}(\xi) \pmod{p}.$$

The same is true for the functions $\mathfrak{F}^{(2)}(\xi), \dots, \mathfrak{F}^{\lambda}(\xi)$ if they are also not decomposable for p.

Now[49] all whole numbers from $(\mathfrak{G})$, and hence also ξ_0, satisfy the congruence

$$\xi^{p^\kappa} - \xi \equiv 0 \pmod{P^{(i)}} \qquad {\scriptstyle (i=1,2,\dots,\lambda)},$$

for each of the λ divisors of degree κ, namely $P^{(1)}, \dots, P^{(\lambda)}$. Hence the expression $(\xi^{p^\kappa} - \xi)$ is divisible modulo p by each of the λ functions $\mathfrak{F}^{(1)}(\xi), \dots \mathfrak{F}^{(\lambda)}(\xi)$, if they are assumed to be irreducible. If those functions are incongruent modulo p, then the function $(\xi^{p^\kappa} - \xi)$ must be divisible by their product, and so it must contain at least λ irreducible factors modulo p of degree κ.[50] If we consider all the

in Dedekind's "Abriß" [**18**]. Hensel sometimes preferred to provide his own proofs rather than searching the literature. See, for example, [**77**] and [**81**].

46 Now we focus on all the irreducible factors of a given degree.

47 The observation is that any polynomial such that $F(\xi_0) \equiv 0 \pmod{P^{(i)}}$ must be divisible (modulo p) by the corresponding irreducible polynomial $\mathfrak{F}^{(i)}$.

48 Since the polynomials all have integer coefficients.

49 The residue field of each of the primes $P^{(i)}$ has p^κ elements, all of which are roots of $x^{p^\kappa} - x$.

50 Hensel has shown, then, that any irreducible polynomial of degree κ is a divisor of $\xi^{p^\kappa} - \xi$ modulo p. In modern terms, adjoining a root of an irreducible polynomial of degree κ to $\mathbb{F}_p$ always gives the same field, namely the splitting field of $\xi^{p^\kappa} - \xi$. In fact, Hensel also needs to know that any polynomial whose degree *divides* κ is a factor modulo p of $\xi^{p^k} - \xi$. For that he quotes his older paper [**75**], which is one of many nineteenth century papers dealing with "higher congruences" that we would describe as being about the theory of finite fields.

irreducible factors modulo p one finds* that $(\xi^{p^\kappa} - \xi)$ is the product of all irreducible polynomials whose degree is equal either to κ or to a divisor of κ and that $(\xi^{p^\kappa} - \xi)$ has exactly

$$(10.) \qquad \bar{g}(\kappa) = \frac{1}{\kappa}\left(p^\kappa - \sum p^{\frac{\kappa}{q}} + \sum p^{\frac{\kappa}{qq'}} - \sum p^{\frac{\kappa}{qq'q''}} + \dots\right)$$

distinct irreducible divisors of degree κ, where $q, q', q'', \dots$ are the distinct prime factors of κ.[51] So if $\lambda > \bar{g}(\kappa)$ then it is not possible for the λ irreducible functions $\mathfrak{F}^{(1)}(\xi), \dots \mathfrak{F}^{\lambda}(\xi)$ to be distinct modulo p. If we now apply this result to all γ of the distinct degrees $\kappa_1, \dots, \kappa_\lambda$ of the prime factors of p, we obtain from theorem (A.):[52]

(B.[53]) Suppose that $p = P_1^{\delta_1} \dots P_h^{\delta_h}$ is the decomposition of a real prime number p in a domain $(\mathfrak{G})$, and that among the h nonequivalent prime factors $P_1, P_2, \dots, P_h$ there are

$$\lambda_1 \text{ of degree } \kappa_1,$$
$$\lambda_2 \text{ of degree } \kappa_2,$$
$$\vdots$$
$$\lambda_\gamma \text{ of degree } \kappa_\gamma.$$

So we can find a number ξ_0 in the domain $(\mathfrak{G})$ that does not satisfy any congruence modulo p of degree less than n only if

$$(11.) \qquad \lambda_1 \leq \bar{g}(\kappa_1),\ \lambda_2 \leq \bar{g}(\kappa_2),\ \dots,\ \lambda_\gamma \leq \bar{g}(\kappa_\gamma)$$

holds (where the γ whole numbers $\bar{g}(\kappa)$ are as in (10.)). If however, even one of these conditions is not met, then *every* number ξ_0 from $(\mathfrak{G})$ satisfies a polynomial congruence modulo p of degree less than n.

It should now be proved that condition (11.) is also sufficient[54] to guarantee that at least one number ξ_0 from $(\mathfrak{G})$ satisfies no congruence modulo p of degree less than n. Let P be one of the h prime divisors of p and let κ be its degree. Then

[51]The point is that we have a polynomial of degree p^κ which is the product of all irreducible polynomials mod p whose degree divides κ. Writing p^κ as the sum of those degrees and using Möbius inversion gives formula (10.) for the total number of distinct irreducible polynomials of degree κ in $\mathbb{F}_p[\xi]$. Notice that Hensel does not use the notation μ or even give a name for that function. An identical formula is also found in Dedekind's *Abriß* **[18]** (page 62 in **[25**, vol. I], but Dedekind does not quote it in the 1878 paper.

[52]In the statement of this theorem Hensel uses "real prime" to refer to a prime in $\mathbb{Z}$. Similarly, he later uses "real integer" for an element of $\mathbb{Z}$. The modern usage is "rational integer" and "rational prime," but we have preserved Hensel's words.

[53]This is just theorem A plus an explicit count of the number of irreducible polynomials in degree κ in $\mathbb{F}_p[\xi]$.

[54]So another proof is beginning here: that the conditions (11.) imply the existence of at least one ξ_0 with the desired property. This is equivalent to Theorem (IV) in Chapter 4, but expressed in terms of congruences rather than index divisors. Hensel's proof is identical to Dedekind's.

*Compare with my paper: Untersuchung der ganzen algebraischen Zahlen eines Gattungsbereiches für einen beliebigen algebraischen Primdivisor; this Journal, Volume 101, pages 140 and 141.

we can choose[55] ξ_0 to satisfy the irreducible polynomial congruence

$$\mathfrak{F}(\xi) \equiv 0 \pmod{P},$$

where $\mathfrak{F}(\xi)$ is one of the $\bar{g}(\kappa)$ irreducible divisors of degree κ of $(\xi^{p^\kappa} - \xi)$.

In[56] the congruence

$$\xi^{p^\kappa} - \xi \equiv \mathfrak{F}(\xi)\Phi(\xi) \pmod{P},$$

both the left and ride side disappear for as many incongruent values of ξ as the degree (namely for the p^κ congruence classes modulo P of numbers in the domain $(\mathfrak{G})$). There must therefore exist a number ξ_0 for which $\mathfrak{F}(\xi_0)$ is divisible by P, because if not the function $\Phi(\xi)$ of degree $(p^\kappa - \kappa)$ would vanish modulo P for p^κ incongruent values of ξ, which is not possible. We can therefore[57] choose ξ_0 to be a root of the chosen irreducible congruence of degree κ

$$\mathfrak{F}(\xi) \equiv 0 \pmod{P}.$$

Now choose[58] h irreducible functions $\mathfrak{F}_1(\xi)$, $\mathfrak{F}_2(\xi)$, ..., $\mathfrak{F}_h(\xi)$, all incongruent modulo p, whose degrees are respectively equal to $\kappa_1, \kappa_2, \dots \kappa_h$, so that each of the $\mathfrak{F}_i(\xi)$ is a divisor of $\xi^{p^{x_i}} - \xi$. Such a system of h functions can only exist when condition (11.) is satisfied. When it is, we can find such a system, since the quantity of irreducible functions incongruent modulo p of degrees $\kappa_1, \dots, \kappa_h$ is larger than the number of functions we need. Let then

$$\xi_0^{(1)}, \xi_0^{(2)}, \dots, \xi_0^{(h)} \tag{12.}$$

be h integers from $(\mathfrak{G})$ chosen so that for each i the integer $\xi_0^{(i)}$ satisfies modulo P_i the congruence

$$\mathfrak{F}_i(\xi_0^{(i)}) \equiv 0 \pmod{P_i} \qquad (i = 1, \dots, h). \tag{12$^\text{a}$.}$$

As we proved above, we can find h such numbers $\xi_0^{(i)}$.

Moreover, we can also assume[59] from the outset that the left side in (12$^\text{a}$.) is not divisible by P_i^2, so that

$$\mathfrak{F}_i(\xi_0^{(i)}) \not\equiv 0 \pmod{P_i^2} \qquad (i = 1, 2, \dots h). \tag{12$^\text{b}$.}$$

This is possible because if for some i

$$\mathfrak{F}_1(\xi_0^{(1)}) \equiv 0 \pmod{P_1^2},$$

[55]This is the lemma to be proved next. Given an irreducible polynomial of degree κ in $\mathbb{F}_p[x]$, we can choose ξ_0 so that it is a root of that polynomial modulo P.

[56]Here starts the proof of the lemma.

[57]Lemma has been proved.

[58]We have shown that for each prime P of degree κ and each irreducible polynomial of degree κ we can find an integer that is a root of that polynomial modulo P. Now we apply this to each of the prime factors of p.

[59]Another little lemma.

one can substitute $\xi_0^{(1)}$ by $\bar{\xi}_0^{(1)} = (\xi_0^{(1)} + \pi_1)$, where π_1 is an integer[60] divisible by P_1 but not P_1^2. Then using[61] *Taylor*'s theorem, we have:

$$\mathfrak{F}_1(\bar{\xi}_0^{(1)}) = \mathfrak{F}_1(\xi_0^{(1)} + \pi_1) = \mathfrak{F}_1(\xi_0^{(1)}) + \pi_1\mathfrak{F}_1'(\xi_0^{(1)}) + \frac{1}{2}\pi_1^2\mathfrak{F}_1''(\xi_0^{(1)}) + \dots.$$

According to our assumptions, the first term as well as the third and all subsequent terms[62] are divisible by P_1^2. But in the second term $\mathfrak{F}_1'(\xi_0^{(1)})$ is not divisible by P_1 (because $\mathfrak{F}_1'(\xi)$ cannot have a common divisor[63] with the irreducible polynomial $\mathfrak{F}_1(\xi)$), we see the following congruence:

$$\mathfrak{F}_1(\bar{\xi}_0^{(1)}) \equiv \pi_1\mathfrak{F}_1'(\xi_0^{(1)}) \not\equiv 0 \pmod{P_1^2}.$$

Thus,[64] the h numbers $\xi_0^{(1)}, \dots, \xi_0^{(\lambda)}$ can be chosen from the beginning in such a way that the h terms

$$\mathfrak{F}_1(\xi_0^{(1)}),\ \mathfrak{F}_2(\xi_0^{(2)}),\ \dots,\ \mathfrak{F}_h(\xi_0^{(h)})$$

are divisible once and only once by the corresponding divisors

$$P_1,\ P_2,\ \dots,\ P_h.$$

If this occurs, then it is also possible to find[65] an algebraic number ξ_0, so that

$$\text{(13.)} \qquad \xi_0 \equiv \xi_0^{(i)} \pmod{P_i^2} \qquad (i = 1, 2, \dots, h).$$

There always exists[66] a number Π_1 in the domain $(\mathfrak{G})$ that is not divisible by P_1, but is divisible by all other divisors of p. We can find another number ϱ_1 so that*

$$\varepsilon_1 = \varrho_1\Pi_1^2 \equiv 1 \pmod{P_1^2}.$$

Then the number ε_1 is divisible by each of the divisors $P_2^2, \dots, P_h^2$, while the remainder of division by P_1^2 is 1. So we take

$$\varepsilon_1,\ \varepsilon_2,\ \dots,\ \varepsilon_h$$

---

[60] So π_1 is a uniformizer at P_1.

[61] Like Dedekind (see Chapter 4, footnote 114), Hensel doesn't discuss the denominators in the Taylor expansion. What is actually needed is a version without denominators of Taylor's theorem for polynomials: express the formal polynomial $f(X+Y)$ as a polynomial in Y with coefficients in $\mathbb{Z}[X]$.

[62] If $p = 2$ it does not seem clear that the third term is divisible by P_1^2. See the previous footnote.

[63] Hensel is assuming, perhaps without noticing it, that he doesn't have to worry about the possibility that $\mathfrak{F}_i'(\xi) \equiv 0$. He is correct because finite fields are perfect.

[64] The lemma has been proved.

[65] In the parallel passage of **[23]** (see Chapter 4, footnote 120), Dedekind simply invokes the Chinese Remainder Theorem from **[21]**; as often, Hensel is going to give a proof.

[66] Here begins the proof of the Chinese Remainder Theorem in this situation. The number Π_1 is a co-uniformizer at p; such numbers played a crucial role in Kummer's cyclotomic theory.

*Set $\varrho_1 = x_1 + \pi_1 y_1$, where π_1 is divisible by P_1 exactly once. Then the algebraic numbers x_1 and y_1 are defined by the system of linear congruences for the Modulus P_1

$$x_1\Pi_1^2 \equiv 1 \pmod{P_1}, \qquad \frac{(x_1 + \pi_1 y_1)\Pi_1^2 - 1}{\pi_1} \equiv 0 \pmod{P_1},$$

which always has a solution.

to be h algebraic integers chosen so that

$$\begin{aligned} \varepsilon_i &\equiv 1 \pmod{P_i^2} \\ \varepsilon_i &\equiv 0 \pmod{P_l^2}, \qquad (i \neq l) \end{aligned}$$

and set

$$\xi_0 = \epsilon_1 \xi^{(1)} + \epsilon_2 \xi_0^{(2)} + \cdots + \epsilon_h \xi_0^{(h)}. \tag{14.}$$

Then ξ_0 satisfies the h conditions (13.).[67] Now from (12^{a}) and (12^{b}), we have

$$\begin{cases} \mathfrak{F}_i(\xi_0) \equiv \mathfrak{F}_i(\xi_0^{(i)}) \equiv 0 \pmod{P_i} \\ \mathfrak{F}_i(\xi_0) \equiv \mathfrak{F}_i(\xi_0^{(i)}) \not\equiv 0 \pmod{P_i^2} \end{cases}. \tag{15.}$$

Moreover, the expression $\mathfrak{F}_i(\xi_0)$ is not divisible by any prime divisors P_l distinct from P_i. (If so, since $\mathfrak{F}_l(\xi_0)$ is certainly divisible by P_l, the irreducible polynomials $\mathfrak{F}_i(\xi)$ and $\mathfrak{F}_l(\xi)$ would have a common divisor, which means[68] they would be congruent to each other, which conflicts with the above assumption.)

Since ξ_0 is chosen according to the condition (14.), for each i this number satisfies the irreducible congruence of degree κ_i

$$\mathfrak{F}_i(\xi_0) \equiv 0 \pmod{P_i}$$

and this expression is not divisible by any other[69] prime factor, which means:[70]

$$p + u.\mathfrak{F}_i(\xi_0) \sim P_i.$$

If this is the case, then we can prove, in exactly the same way as was done for the fundamental form w_0 in §3 of the previous work,[71] that the congruence of degree n

$$\mathfrak{F}_1^{\delta_1}(\xi)\mathfrak{F}_2^{\delta_2}(\xi)\ldots\mathfrak{F}_h^{\delta_h}(\xi) \equiv 0 \pmod{p},$$

is the smallest[72] that ξ_0 satisfies, which means it is the element we need for the converse[73] of Theorem (B.).

Recalling that the degree κ of a prime divisor P of p coincides with the degree (as a polynomial in w) of the corresponding irreducible factor of $\mathfrak{F}(w)$ from (2.), we can restate the result without using[74] the prime factorization of p, in the following manner:

---

[67] So we have proved the Chinese Remainder Theorem.

[68] Because they are irreducible.

[69] Hensel means "by any other prime factor of p."

[70] Here $\sim$ denotes equivalence of divisors in Kronecker's sense. We would say that P_i is the greatest common divisor of p and $\mathfrak{F}_i(\xi_0)$.

[71] Again, this is [**79**].

[72] I.e., is of the smallest degree.

[73] So this concludes the proof of the converse. See Chapter 4, footnote 129 for a numerical example.

[74] The idea, of course, is that in general it is hard to find the factorization of p, and especially so when the condition in the theorem holds. Alas, factoring the fundamental equation is hard as well. Hensel will address this in the next section.

If

$$\mathfrak{F}(w) \equiv \mathfrak{F}_1{}^{\delta_1}(w) \ldots \mathfrak{F}_h^{\delta_h}(w) \pmod{p}$$

is the decomposition of the fundamental equation of a genus $(\mathfrak{G})$ into its irreducible factors modulo p, and if the numbers

$$\lambda_1, \lambda_2, \ldots, \lambda_\gamma,$$

indicate how many of the h factors $\mathfrak{F}_1(w), \ldots, \mathfrak{F}_h(w)$ have corresponding degree

(C.) $$\kappa_1, \kappa_2, \ldots, \kappa_\gamma,$$

then *all* values ξ_0 of domain satisfy a congruence of degree lower than n modulo p if and only one of the γ conditions

$$\lambda_i > \bar{g}(\kappa_i)(i = 1, 2, \ldots, \gamma)$$

is satisfied. Here the term $\bar{g}(\kappa)$ is

$$\bar{g}(\kappa) = \frac{1}{\kappa}(p^\kappa - \sum p^{\frac{\kappa}{q}} + \sum p^{\frac{\kappa}{qq'}} - \ldots),$$

and $q, q', \ldots$ are the distinct prime factors of the number κ.

Now[75] take ξ_0 to be some number in the domain $(\mathfrak{G})$; listing the first n powers of ξ_0 in terms of the fundamental system[76] $\xi_1, \ldots, \xi_n$, we get n equations with rational integer coefficients

$$(16.)\qquad \begin{cases} 1 = a_{10}\xi_1 + \cdots + a_{n0}\xi_n, \\ \xi_0 = a_{11}\xi_1 + \cdots + a_{n1}\xi_n, \\ \vdots \\ \xi_0^{n-1} = a_{1,n-1}\xi_1 + \cdots + a_{n,n-1}\xi_n. \end{cases}$$

Now ξ_0 satisfies a polynomial congruence modulo p of degree less than n if and only if the determinant

$$|a_{ik}| \qquad \begin{matrix} (i = 1, \ldots, n) \\ (k = i, 1, \ldots, n-1) \end{matrix}$$

of the n linear equations (16.) is divisible by p. This is because[77] only in this case can we find n numbers $A_0, A_1, \ldots, A_{n-1}$ (not all divisible by p), such that the sum of the first equation multiplied with A_0, the second with A_1, ..., and the last with A_{n-1}, gives

$$A_0 + A_1\xi_0 + \cdots + A_{n-1}\xi_0^{n-1} \equiv 0 \pmod{p}$$

by making the coefficients of $\xi_1, \ldots, \xi_n$ on the right side of the equation all[78] divisible by p. When we form the n systems of equations from (16.) and consider the n

[75]The next few paragraphs relate the fact that ξ_0 satisfies a congruence of degree less than n to its index.

[76]Recall that "fundamental system" is Kronecker's name for an integral basis.

[77]A bit of linear algebra modulo p: we want the system $A[\xi] = [0]$ to have a nontrivial solution mod p, which requires the determinant to be zero mod p.

[78]Since $\xi_1, \ldots, \xi_n$ is an integral basis, the only way an algebraic integer will be divisible by p is by having all the coefficients divisible by p.

conjugate[79] domains to $\mathfrak{G}$, then we see the validity of the equation

$$\mathfrak{D}(\xi_0) = |a_{ik}|^2.D,$$

where $\mathfrak{D}(\xi_0)$ is the discriminant of the equation for ξ_0 and D is the discriminant of the genus $(\mathfrak{G})$.

Each equation discriminant consists, then, of two essentially different parts: on the one hand, the domain discriminant D, which is the same in all discriminants, and on the other the squared determinant $|a_{ik}|^2$, which is dependent on the choice of ξ_0. For this reason, *Kronecker* called the first the essential and the second the inessential divisor of the discriminant $\mathfrak{D}(\xi_0)$. A prime p is contained in $|a_{ik}|^2$ (and so is an inessential divisor of the discriminant $\mathfrak{D}(\xi_0)$) if and only if ξ_0 satisfies a congruence modulo p of degree less than n.[80] From our Theorem (C.), it follows now that the first part $|a_{ik}|^2$ of the discriminant (although it depends on ξ_0) can contain factors which remain the same whatever ξ_0 is chosen, and so cannot be removed by an appropriate choice of ξ_0. These "common inessential divisors" of all equation discriminants of a domain are the primes p (and only these) for which every number ξ_0 of the domain satisfies a congruence of lower than n-th degree. By applying Theorem (C.) we get the following:

(D.) If

$$F(w) \equiv \mathfrak{F}_1^{\delta_1}(w) \dots \mathfrak{F}_h^{\delta_h}(w) \pmod{p}$$

is the decomposition of the fundamental equation of a genus $(\mathfrak{G})$ into its irreducible factors modulo p, and if the numbers $\lambda_1, \dots, \lambda_\gamma$ indicate how many factors have degree $\kappa_1, \dots, \kappa_\gamma$ as polynomials in w, then p is a common inessential divisor of all equation discriminants $\mathfrak{D}(\xi_0)$ from $(\mathfrak{G})$ if and only if at least one of the γ conditions

$$\lambda_1 > \bar{g}(\kappa_1), \dots, \lambda_\gamma > \bar{g}(\kappa_\gamma)$$

is satisfied.

§2

The result from the previous section can also be expressed in another form that is remarkable in that to apply it we do not need to know the decomposition of p within the domain $(\mathfrak{G})$ or the factorization of the fundamental equation modulo p.[81]

---

[79]This is basically matrix multiplication: the matrix whose columns are the powers of ξ_0 and its conjugates is equal to $[a_{ik}]$ times the matrix whose columns are the integral basis and its conjugates. Hensel thinks of n conjugate domains rather than doing the computation in a normal closure.

[80]Dedekind **[23]** called the (absolute value of the) determinant $|a_{ik}|$ the *index* of the algebraic integer ξ_0. Hensel seems to be content not to have a name for it.

[81]The obvious objection to the criterion above is that in order to use it we need to know how p factors, or, equivalently, we need to be able to factor the fundamental equation modulo p. This can be hard to do in general. But in fact one does not need to know the full factorization. It suffices to know, for each $\kappa = 1, 2, \dots, n$, the number λ_κ of distinct primes of degree κ occurring in the factorization of p. Then we can compare that with the number $\bar{g}(\kappa)$ of irreducible polynomials of degree κ in $\mathbb{F}_p[x]$ to determine whether p is a common inessential discriminant divisor. The goal of this section is to present a way of computing λ_κ without finding the full factorization of p. From this point on Hensel is venturing beyond Dedekind's 1878 paper **[23]** translated in Chapter 4.

Let P be an arbitrary prime factor of p and let κ be its [residual] degree. Now if

$$w_0 = u_1\xi_1 + \cdots + u_n\xi_n$$

is a fundamental form[82] for the genus domain $(\mathfrak{G})$, we set

$$w_h = u_1\xi_1^{p^h} + \cdots + u_n\xi_n^{p^h} \qquad (h = 0, 1, \ldots).$$

We know from the previous work[83] (page 65), that only the first κ of these infinitely many forms[84]

$$w_0, w_1, \ldots, w_{\kappa-1}$$

are distinct modulo P (for indeterminate $u_1, \ldots, u_n$). In fact, $w_\kappa \equiv w_0$, $w_{\kappa+1} \equiv w_1, \ldots$, and in general

$$w_{h+\kappa} \equiv w_h \pmod{P}.$$

From this it follows that the linear form[85]

$$w_\nu - w_0 = u_1(\xi_1^{p^\nu} - \xi_1) + u_2(\xi_2^{p^\nu} - \xi_2) + \cdots + u_n(\xi_n^{p^\nu} - \xi_n)$$

is divisible by the prime divisor P if and only if the number ν is a multiple of the degree κ. If this is the case, then the following simple considerations show that P is only contained once in that linear form.[86]

If, first, P is a multiple divisor of p, then P^2 is contained in p and we know from *Fermat*'s [Little] theorem, that for all integer values of $u_1, \ldots, u_n$ the congruence

$$w_\nu \equiv w_0^{p^\nu} \pmod{p}$$

holds. The same is also fulfilled a fortiori modulo the divisor P^2 of p. If it were true that even for indeterminate $u_1, \ldots, u_n$ we had P^2 dividing the linear form $w_\nu - w_0$, then *all* numbers $\xi_0 = a_1\xi_1 + \cdots + a_n\xi_n$ in $(\mathfrak{G})$ would satisfy the congruence

$$\xi_0^{p^\nu} - \xi_0 \equiv 0 \pmod{P^2}.$$

That this is not the case[87] can be easily seen if we consider, for example, $\xi_0 = \pi$, where π is divisible by P, but not by P^2; in this case, the left side of the congruence reduces modulo P^2 to $(-\pi)$, since π^{p^ν} is clearly divisible by P^2.

---

[82] That is, $\{\xi_1, \xi_2, \ldots, \xi_n\}$ is an integral basis and the u_i are indeterminates. Everything that follows is dependent on the choice of an integral basis.

[83] Namely **[79]**.

[84] The forms w_h are lifts of the images of w_0 under the Frobenius automorphism modulo P, which is of order κ. This gives the congruence claims that follow immediately. Hensel does not have any of this language at his disposal, of course.

[85] This is the key object for this section. In Dedekindian terms, we are considering the ideal I_ν generated by the elements $\xi_i^{p^\nu} - \xi_i$. As noted above, these ideals depend on the choice of integral basis.

[86] The claim is that when $\kappa|\nu$ the ideal I_ν is divisible exactly once by each prime of degree κ appearing in the factorization of p. It is clear that any such P divides $w_\nu - w_0$, so the key thing to prove is that P^2 does not. The next two paragraphs appear to provide a proof of this claim. In fact, however, what they show is that the forms $w_\nu - w_0$ may need to be modified so that this is true. Hensel phrases this as modifying the integral basis, but after the modification he suggests the list $\xi_1, \xi_2, \ldots, \xi_n$ is no longer an integral basis.

The "proof" is divided into two cases: when P^2 divides p and when it does not. In the first case, we have an actual proof. In the second a change to the integral basis is needed. Hensel is assuming we do not know the factorization of p, however, so he will make the modification in any case.

[87] So when P^2 divides p the form $w_\nu - w$ is never divisible by P^2. The argument involves choosing a uniformizer at p.

Next, if P divides p only once and if the linear form $w_\nu - w_0$ for indeterminate $(u_1, \dots, u_n)$ is divisible by P^2, then we must have

$$\xi_i^{p^\nu} - \xi_i \equiv 0 \pmod{P^2} \qquad (i = 1, 2, \dots, n)$$

for the n elements of the fundamental system, which as can easily be seen is generally[88] not the case. For if the system $(\xi_1, \dots, \xi_n)$ did have this property, we could modify it, without changing its character[89] modulo p, so that this exception does not occur. It suffices to replace one of the n elements ξ_i by $\xi_i + p$. We know from the binomial theorem that

$$(\xi_i + p)^{p^\nu} - (\xi_i + p) \equiv (\xi_i^{p^\nu} - \xi_i) - p \equiv -p \not\equiv 0 \pmod{P^2},$$

because all other terms are divisible by p^2 and so by P^2. The easiest way of avoiding the occurrence of this case, without knowing the prime factors of p, is, as is always possible, to assume that the first element of the fundamental system a priori equals one, and then instead introduce $\xi_1 = 1 + p$, which does not change the character of the fundamental system modulo p at all.[90] Then this exception cannot occur for any prime factor of p because

$$\xi_1^{p^\nu} - \xi_1 = (1+p)^{p^\nu} - (1+p) \equiv -p \pmod{p^2}.$$

In this case $w_\nu - w_0$ will not contain any prime divisors from p more than once.

The result of these quick observations is summarized in the following theorem.[91]

> When $u_1, \dots, u_n$ are indeterminates, the linear form
>
> $$w_\nu - w_0 = u_1(\xi_1^{p^\nu} - \xi_1) + \cdots + u_n(\xi_n^{p^\nu} - \xi_n),$$
>
> which is of degree p^ν with respect to the elements of the fundamental system $\xi_1, \dots, \xi_n$, is divisible by the product of all distinct prime divisors of p whose degree κ is an exact divisor of ν, and contains each of these exactly once.

---

[88] In the unramified case it is indeed possible for $\xi_i^{p^\nu} - \xi_i$ to be divisible by P^2. A simple example is to take $K = \mathbb{Q}(\sqrt{3})$ with $\xi_1 = 1$ and $\xi_2 = \sqrt{3}$. Let $p = 11$, Then $\xi_1^{11} - \xi_1 = 0$ and

$$\xi_2^{11} - \xi_2 = (\sqrt{3})^{11} - \sqrt{3} = 242\sqrt{3} = 2 \cdot 11^2\sqrt{3}.$$

If, as Hensel suggests, we instead use $\xi_1 = 12$, then it works, since $12^{11} - 12$ is divisible by 11 only once.

The most dramatic example is the cyclotomic field generated by an ℓ-th root of unity when $p \equiv 1 \pmod{\ell}$. If we take the standard integral basis, then $w_\nu = w_0$ for all ν, and $w_\nu - w_0 = 0$.

For a cubic example, let $K = \mathbb{Q}(\alpha)$ with $\alpha^3 - 6\alpha^2 - 9\alpha - 1 = 0$ (number field 3.3.3969.2 in [**109**]). The integral basis is $(1, \alpha, \alpha^2)$ and when $p = 5$ both $\alpha^5 - \alpha$ and $\alpha^{10} - \alpha^2$ turn out to be divisible by the square of one of the primes dividing 5.

[89] Hensel is correct that this does not change the reduction mod p of the ξ_i, but of course they may no longer be an integral basis. This comment makes it clear that Hensel really wants to work "locally at p," though he does not really have the language to say that.

[90] So Hensel is telling us to always assume $\xi_1 = 1 + p$, which guarantees that the ξ_i are no longer an integral basis, since $1 = \frac{1}{p+1}\xi_1$ is an integer. This is not mentioned, however, in the statements of the theorems that follow.

The whole argument is in fact (perhaps inadvertently) local: the ξ_i are only an integral basis at p, and all the divisibility proofs below consider only the primes above p.

[91] In terms of ideals, the theorem can be stated thus. Let $1, \xi_2, \dots, \xi_n$ be an integral basis, and let $\xi_1 = 1 + p$. Let I_ν be the ideal generated by $\xi_i^{p^\nu} - \xi_i$, $i = 1, 2, \dots, n$. Then I_ν is divisible exactly once by each prime ideal of degree dividing ν that appears in the factorization of p.

Note that Hensel does not say (or prove) that his linear form has no divisors prime to p. Later, he seems to assume that this is the case, but it is not true in general.

From this theorem we can draw an interesting conclusion,[92] which is of importance for a subsequent[93] investigation:

If the prime number p contains prime factors which are pairwise distinct, so that

$$p \sim P_1 P_2 \dots P_h,$$

then we can always find a linear form $w_\nu - w_0$ divisible by the prime p, or, equivalently, such that the n congruences:

$$\xi_i^{p^\nu} \equiv \xi_i \pmod{p} \qquad (i = 1, 2, \dots, n)$$

are all satisfied. Given the theorem, we only need to choose ν to be the least common multiple of the h degrees $\kappa_1, \dots, \kappa_h$ of the prime divisors $P_1, \dots, P_h$.

On the other hand, if p contains a prime divisor more than once, then none of the linear factors[94] $w_\nu - w_0$ is divisible by p, because none of those differences can contain a multiple factor of p more than once. Thus we have the following theorem:

> The prime p decomposes in the domain $(\mathfrak{G})$ into a product of distinct prime divisors if and only if at least one of the linear forms $w_\nu - w_0$ is divisible by p.

Since the number p is a divisor of the genus discriminant[95] when and only when it contains at least one multiple prime factor, we can state as a corollary of the previous result the following theorem:[96]

> The prime p is contained in the discriminant of the domain $(\mathfrak{G})$ if and only if one of the linear forms $(w_\nu - w_0)$ is divisible by a fractional power of p, but not divisible by p itself.

Now let κ be an arbitrary whole number. We will form the product:

$$F_\kappa(w_0) = \frac{(w_\kappa - w_0)\prod(w_{\frac{\kappa}{qq'}} - w_0)\prod(w_{\frac{\kappa}{qq'q''q'''}} - w_0)\dots}{\prod(w_{\frac{\kappa}{q}} - w_0)\prod(w_{\frac{\kappa}{qq'q''}} - w_0)\dots},$$

where q, q', q'', … are the distinct prime factors of κ. More simply,

$$F_\kappa(w_0) = \prod_{d|\kappa}(w_d - w_0)^{\varepsilon_d},$$

where $\varepsilon_d = \pm 1$, according to whether the divisor of κ complementary to d is a product of an even or odd number of *distinct* prime factors q, q', q'' … of κ, and where $\epsilon_\delta = 0$ when the ratio $\frac{\kappa}{d}$ contains repeated prime factors.[97] This quotient is

---

[92]Hensel will use the theorem to derive a criterion for p to be ramified in $(\mathfrak{G})$. The point is that p is unramified if and only if it divides one of the ideals I_ν. Note that he knows that this is equivalent to p not dividing the field discriminant.

[93]As Petri notes in [**117**, 2.4], it is unclear which "subsequent investigation" (folgende Untersuchung) Hensel has in mind.

[94]Sic, but he wants to say "linear forms."

[95]"Gattungsdiscriminante."

[96]We have translated the theorem as Hensel states it, but the statement is incorrect. What he had proved is that p is unramified if and only if it divides one of the forms $w_\nu - w$ (equivalently, one of the ideals I_ν). The negation would then say that p is ramified if and only if it divides none of them. So "one" should be "none." See the numerical example in Section 6.5.1.

[97]In modern terms $\varepsilon_d = \mu(\kappa/d)$, where μ is the Möbius function. It apparently was introduced by Möbius in 1832, but was clearly not yet part of the standard toolkit.

a rational function[98] of the elements $(\xi_1, \ldots, \xi_n)$ of the fundamental system of $(\mathfrak{G})$. Its dimension[99] with respect to these elements is

$$g(\kappa) = p^\kappa - \sum_q p^{\frac{\kappa}{q}} + \sum_q p^{\frac{\kappa}{qq'}} - \cdots = \sum_{d|\kappa} \varepsilon_d p^d.$$

We immediately recognize that it [namely, $F_\kappa(w_0)$] is equivalent to[100] the product of all distinct prime divisors of p whose degree is exactly equal to κ. If[101] $\bar{P}$ is a prime divisor of p whose degree $\bar{\kappa}$ is not a divisor of κ, then $\bar{P}$ is contained in neither the numerator or denominator of $F_\kappa(w_0)$. If[102] however, $\bar{\kappa}$ is a divisor of κ then one shows exactly as in the corresponding question[103] in the theory of cyclotomic equations, that $\bar{P}$ occurs in the denominator just as often as it occurs in the numerator of $F_\kappa(w_0)$. If[104] $\bar{\kappa} = \kappa$ then $\bar{P}$ is contained once and only once in the numerator of the $F_\kappa(w_0)$, in the linear form $(w_\kappa - w_0)$, and thus our claim is proved.[105]

If then $P^{(1)}, P^{(2)}, \ldots P^{(\lambda_\kappa)}$ are all the distinct prime factors of p whose degree is equal to κ, then $F_\kappa(w_0)$ is equal[106] to their product. This means, we have an equivalence:

$$F_\kappa(w_0) = \prod_{d|\kappa} (w_d - w_0)^{\epsilon_d} \sim P^{(1)} P^{(2)} \ldots P^{(\lambda_\kappa)}.$$

Taking norms,[107] it follows[108] that

$$N(F_\kappa(w)) = p^{\kappa\lambda_\kappa} = p^{L_\kappa}$$

[So we get[109]]

[98]In terms of ideals, the $F_\kappa(w_0)$ correspond to fractional ideals $F_\kappa = \prod_{d|\kappa} I_d^{\varepsilon_d}$, where the I_d are as above. At this point Hensel's choice of notation emphasizes the dependence of F_κ on the fundamental form w_0, i.e., on the choice of integral basis. Hensel's argument shows that in fact F_κ is integral at p, but as usual does not consider any other divisors.

[99]Hensel means the degree of the rational function in the symbols ξ_i.

[100]The claim is that this rational function is equivalent as a divisor, in the sense of Kronecker, to a product of prime divisors. In Dedekind's terms, Hensel is saying that the ideal F_κ is the product of these primes. Strictly speaking this is incorrect, since Hensel ignores completely the primes that do not divide p, but it is correct at p. In other words, the argument continues to be local.

[101]First, divisors of degree not dividing κ do not divide any of the forms that make up $F_\kappa(w_0)$.

[102]Next, divisors of degree dividing κ but unequal to κ cancel out. This is the point of the complicated quotient.

[103]The reference is to the formula for the n-th cyclotomic polynomial $\Phi_n(x)$ in terms of the polynomials $x^d - 1$ for d dividing n. See, for example, [**101**, p. 285], where the notation ε_d is also used.

[104]Divisors of degree κ occur exactly once.

[105]Paragraph break inserted here to improve readability. There is no attempt to deal with primes that are not divisors of p. See the numerical example in Section 6.5.1.

[106]As a divisor in Kronecker's sense.

[107]The norm of a prime of degree κ is of course p^κ. The equation is not actually true, since Hensel is silently ignoring primes that are not divisors of p.

[108]The equation effectively defines the number L_κ. Numerical examples show (see Section 6.5.1) that the norm need not be a power of p, so we should take L_κ as the p-adic valuation of the norm instead.

[109]We have left the statement of both theorems in Hensel's terms, "dimension" and "degree" unchanged. The first means the degree of the rational function, while the second means the residual degree of the corresponding divisor. The "dimension" is just $g(\kappa)$, which we can easily compute in any case.

The form $F_\kappa(w)$, which has dimension $g(\kappa)$ with respect to $\xi_1, \dots, \xi_n$, has degree $\kappa\lambda_\kappa$, where λ_κ is the number of distinct prime factors of p of degree κ. If no prime factor of degree κ exists, then $\lambda_\kappa = 0$.

If we construct the n forms

$$F_\kappa(w_0) \qquad (\kappa = 1, 2, \dots, n),$$

we know that their degree[110] L_κ equals $\kappa\lambda_\kappa$; from the previously proven theorem (D.), the prime p is an inessential divisor of all equation discriminants $\mathfrak{D}(\xi_0)$ from $(\mathfrak{G})$ when at least one of the inequalities

$$\lambda_\kappa > \bar{g}(\kappa) = \frac{1}{\kappa} g(\kappa) \qquad \text{or} \qquad \kappa\lambda_\kappa > g(\kappa)$$

holds.[111] Now since $\kappa\lambda_\kappa$ is the degree of the form $F_\kappa(w_0)$ and $g(\kappa)$ is the dimension with respect to $\xi_1, \dots, \xi_n$, we can state the previously found result in a more elegant and simple[112] form:

The prime p is a common inessential divisor of the equation discriminants $\mathfrak{D}(\xi_0)$ of the ring of integers $(\mathfrak{G})$ if among the forms

$$F_\kappa(w_0) = \prod_{d|\kappa} (w_d - w_0)^{\epsilon_d} \qquad (\kappa = 1, \dots, n)$$

at least one exists whose dimension with respect to the elements $\xi_1, \dots, \xi_n$ of the fundamental system is smaller than the degree, that is than the exponent L_κ of p in the equation

$$N(F_\kappa(w)) = p^{L_\kappa}.$$

§3

The question of common inessential discriminant divisors can now be handled in an entirely different fashion, leading to an entirely different criterion[113] for them to occur.

Let

$$\xi_1^{(0)}, \dots, \xi_n^{(0)}$$

110 "Ordnungszahlen." L_κ is actually the p-adic valuation of the norm. It follows from Hensel's argument that the p-part of the norm of F_κ is integral and independent of the choice of integral basis.

111 For a numerical example of Hensel's results, see Section 6.5.1.

112 It is unclear to us why Hensel prefers this form of the theorem. In any case, the content is the same. Note that the final equation in the theorem only gives the p-part of the norm.

113 The results in this section had all been announced in the introduction to Hensel's thesis; see theorems Q and R in [**74**, p. 11].

be a fundamental system[114] for the genus $(\mathfrak{G})$ and let

$$\begin{aligned} w^{(0)} &= u_1\xi_1^{(0)} + \cdots + u_n\xi_n^{(0)}, \\ w^{(1)} &= u_1\xi_1^{(1)} + \cdots + u_n\xi_n^{(1)}, \\ &\vdots \\ w^{(n-1)} &= u_1\xi_1^{(n-1)} + \cdots + u_n\xi_n^{(n-1)} \end{aligned}$$

be the fundamental forms for the genus $(\mathfrak{G})$ and its conjugates. Then the discriminant of the fundamental equation is

$$D = \prod_{a \neq \beta} (w^{(a)} - w^{(\beta)}) \qquad (a, \beta = 0, 1, \ldots, n-1).$$

This is a homogeneous function of $u_1, \ldots, u_n$ with integer coefficients. The greatest common divisor of all of these coefficients is a whole number, which I have shown in the previous work[115] (page 78) agrees with the genus discriminant, which is to say, with the square of the determinant

$$|\xi_i^{(\kappa)}|^2 \qquad \begin{matrix} (i = 1, 2, \ldots, n) \\ (\kappa = 0, 1, \ldots, n-1) \end{matrix}.$$

So we have[116]

$$D(u_1, \ldots, u_n) = \Delta^2(u_1, \ldots, u_n).|\xi_i^{(\kappa)}|^2, \tag{3.}$$

where $\Delta(u_1, \ldots, u_n)$ is a homogeneous function[117] of $u_1, \ldots u_n$, whose dimension is clearly equal to $\frac{n(n-1)}{2}$, and whose coefficients no longer have any common divisors, which is to say $\Delta(u_1, \ldots, u_n)$ is a primitive polynomial function in $u_1, \ldots, u_n$.

Now suppose that instead of $w^{(0)}$ we choose a number from the domain

$$\xi^{(0)} = a_1\xi_1^{(0)} + \cdots + a_n\xi_n^{(0)}$$

together with its conjugates. Then we obtain its equation discriminant if in (3.) we substitute the indeterminates $u_1, \ldots, u_n$ by the whole numbers $a_1, \ldots, a_n$. The prime p is an inessential divisor of the discriminant if and only if it is contained in $\Delta(a_1, \ldots, a_n)$. So we have the following theorem:[118]

> The prime p is a common inessential equation discriminant divisor for the genus $(\mathfrak{G})$ if and only if the primitive form
>
> $$\Delta(u_1, \ldots, u_n)$$
>
> is divisible by p for all integer values of the indeterminates $u_1, \ldots, u_n$.

---

[114] I.e., an integral basis. The superscripts (0) have been added because Hensel is about to consider conjugates.

[115] As usual, this is **[79]**.

[116] There are no equations in this section marked (1.) or (2.).

[117] In fact Δ is the "index form," i.e., in computes the index of $\mathbb{Z}[\xi^{(0)}]$ in the ring of integers. In the examples given by Dedekind (Chapters 3 and 4), he computes it explicitly.

[118] This is known as Hensel's criterion for common inessential discriminant divisors. The idea is to compute the index form and then check that its values are always divisible by p. Hensel will state it first, then give an explicit way to test a form to see if all its values are indeed divisible by p, then summarize the whole thing into a theorem.

The question of when a polynomial form has values divisible by p for all integer values of the indeterminates is fully answered by the following theorem:

> A form $U(u_1, \ldots, u_n)$ has value divisible by a prime p for all integer values of the indeterminates if it contains the module system:
>
> (4.) $$(p; u_1^p - u_1, \ldots, u_n^p - u_n);$$
>
> that is, when U can be written in the form
>
> (4ª.) $$U(u_1, \ldots, u_n) = pU_0 + (u_1^p - u_1)U_1 + \cdots + (u_n^p - u_n)U_n,$$
>
> where $U_0, U_1, \ldots, U_n$ are integral polynomials in $u_1, \ldots, u_n$.

This theorem can most easily be proved[119] through induction.[120] It is obviously true when no variable is present; we now assume that it is proved for the case of $n-1$ variables $(u_2, \ldots, u_n)$ and prove it for n variables. If the form $U(u_1, \ldots, u_n)$ has degree higher that $p-1$ in u_1, then it can be reduced modulo $u_1^p - u_1$ to another form $\overline{U}(u_1, \ldots, u_n)$, whose degree in u_1 is at most equal to $p-1$, since clearly

$$u_1^p \equiv u_1,\ u_1^{p+1} \equiv u_1^2,\ \ldots,\ u_1^{p+i} \equiv u_1^{i+1} \pmod{(u_1^p - u_1)}.$$

If we write the function so obtained according to the powers of u_1, we get a congruence

(5.) $$U(u_1, \ldots, u_n) \equiv \overline{U} = \overline{U}_0 u_1^{p-1} + \overline{U}_1 u_1^{p-2} + \cdots + \overline{U}_{p-1} \pmod{(u_1^p - u_1)},$$

and the function $\overline{U}(u_1, \ldots, u_n)$ will be divisible by the prime p for every integer system of values of $(u_1, \ldots, u_n)$ when the same is also the case for $U(u_1, \ldots, u_n)$ and conversely, since they differ by a multiple of $u_1^p - u_1$, which is a multiple of p for every integer value of u_1 according to *Fermat*'s theorem.

If we now give $u_2, \ldots, u_n$ any system of integer values $a_2, \ldots, a_n$, [the coefficients] $\overline{U}_0, \ldots \overline{U}_{p-1}$ become equal to integers $A_0, \ldots, A_{p-1}$. The resulting expression for $\overline{U}$,

$$A_o u_1^{p-1} + A_1 u_1^{p-2} + \cdots + A_{p-1},$$

must be divisible by p when we let u_1 be equal to each of the p incongruent numbers $0, 1, \ldots, p-1$. But an expression of degree $p-1$ can only vanish modulo p for p incongruent values of u_1 if *all* its coefficients[121] are divisible by p. So it follows that $\overline{U}(u_1, \ldots, u_n)$ is only divisible by p for all integer value systems if the same is true for the p coefficients $\overline{U}_0(u_2, \ldots, u_n), \ldots, \overline{U}_{p-1}(u_2, \ldots, u_n)$, which are functions only of $u_2, \ldots, u_n$. If this is the case, then, according to the assumption made above, all of these coefficients contain the divisor system $p; u_2^p - u_2, \ldots, u_n^p - u_n$. The same now follows for the whole form $\overline{U}(u_1, u_2, \ldots u_n)$, and from equation (5) it follows that the function being investigated contains the divisor system

$$(p; u_1^p - u_1, \ldots, u_n^p - u_n),$$

since it differs from the previous by only a multiple of $(u_1^p - u_1)$; and thus the theorem is proved.

With the help of this theorem we now have the following criterion for the occurrence of a common inessential discriminant divisor:

---

[119] The proof is in this and the next paragraph.

[120] In fact the proof is constructive: take the form, divide it by $u_1^p - u_1$, look at the coefficients of the resulting polynomial in u_1, rinse, repeat.

[121] I.e., a polynomial of degree $p-1$ cannot have p roots, even modulo p.

> If
> $$\Delta^2(u_1, \dots, u_n)$$
> is the discriminant of the fundamental equation of the domain $(\mathfrak{G})$ with its numerical factor removed, then the prime p is a common inessential divisor of the equation discriminants for $(\mathfrak{G})$ if and only if the primitive polynomial form $\Delta(u_1, \dots, u_n)$ of degree $\frac{1}{2}n(n-1)$ contains the divisor system
> $$(p; u_1^p - u_1, \dots, u_n^p - u_n),$$
> (in the sense of *Kronecker*'s Festschrift), that is, if there is an equation
> $$\Delta(u_1, \dots, u_n) = U_0 p + U_1(u_1^p - u_1) + \dots + U_n(u_n^p - u_n),$$
> where $U_0, U_1, \dots, U_n$ are polynomials in $u_1, \dots, u_n$.

In his Festschrift for the anniversary of *E. E. Kummer*'s doctorate, *L. Kronecker* mentioned[122] the possibility of such common inessential discriminant divisors, adding that this occurs, for example, when the primitive form $\Delta(u_1, \dots, u_n)$ can be expressed as a homogeneous polynomial function of $(u_i^p - u_i)$. The preceding simple observations show us[123] that this can happen *only* in this case, as long as we add a term of the form $pU_0(u_1, \dots, u_n)$ to every expression.[124] But to complete the proof we required the result from the previous paper (p. 78) that the form $\Delta(u_1, \dots, u_n)$ is primitive, which means that the discriminant $D(u_1 \dots, u_n)$ of the fundamental equation contains no numerical divisor beyond the genus discriminant.

If we know the primitive form $\Delta(u_1, \dots, u_n)$, then it will be very easy to determine whether or not p is a common inessential discriminant divisor in $(\mathfrak{G})$. We first reduce the coefficients of this form to their smallest remainder modulo p and all exponents of $u_1, \dots, u_n$ larger than $p-1$ to their smallest remainder modulo $p-1$; then p is a common inessential divisor of the discriminants of $(\mathfrak{G})$ if and only if the resulting form is identical to 0.

To illustrate this way of handling the problem, I will now give the following simple example,[125] which was considered from a different point of view in my doctoral dissertation.

> Let ν be an arbitrary real prime of the form $3\mu + 1$ and $\mathfrak{G}_3(\epsilon)$ be the genus generated by the three μ-fold periods $\epsilon_1, \epsilon_2, \epsilon_3$ of the νth roots of unity. We want to find the common inessential discriminant divisors of this genus.

---

[122] In [**97**, §25]. See Chapter 5

[123] Hensel is saying that Kronecker observed that this condition was sufficient but that what he has added is that it is also necessary, since the content of the discriminant form is exactly the field discriminant.

[124] Since the index form is a homogeneous polynomial in the u_i of degree $n(n-1)/2$, it is an immediate corollary of Hensel's result that if p is a CIDD then $p \leq n(n-1)/2$. Hensel pointed this out in [**82**, p. 291]. Using the results from §1, this can be improved to $p < n$; the usual reference is [**145**], but Bauer already took this as known in [**8**].

[125] In modern language, Hensel looks at the cyclotomic field corresponding to a prime number $\nu = 3\mu + 1$. This has a unique cyclic cubic subfield $\mathfrak{G}_3(\epsilon)$ generated by Gaussian periods $\epsilon_1, \epsilon_2, \epsilon_3$. In his thesis, Hensel found a sufficient condition for such fields (and more general versions of them) to have common inessential discriminant divisors. Since these fields have degree 3, the only prime that can be a CIDD is $p = 2$.

The three periods $\epsilon_1, \epsilon_2, \epsilon_3$ form a fundamental system for the genus[126] $\mathfrak{G}_3(\epsilon)$. Let

$$w_1 = u_1\epsilon_1 + u_2\epsilon_2 + u_3\epsilon_3,$$
$$w_2 = u_1\epsilon_2 + u_2\epsilon_3 + u_3\epsilon_1,$$
$$w_3 = u_1\epsilon_3 + u_2\epsilon_1 + u_3\epsilon_2,$$

so that the product $(w_1 - w_2)(w_2 - w_3)(w_3 - w_1)$ will be the square root of the discriminant of the fundamental equation. By using the known expressions for the resolvent of the cubic period equation, we get without difficulty the expression

$$\prod(w_i - w_{i+1}) = -\nu(\alpha\Delta_1 + \beta\Delta_2).$$

Here α and β are the integers which occur in the decomposition of ν into its prime factors in the genus of the third roots of unity, so that

$$(6.)\qquad \nu = (\alpha + 3\beta\rho)(\alpha + 3\beta\rho^2) = \alpha^2 - 3\alpha\beta + 9\beta^2 \qquad (\rho^2 + \rho + 1 = 0),$$

and Δ_1 and Δ_2 are the primitive forms:

$$\Delta_1 = (u_1 - u_2)(u_2 - u_3)(u_3 - u_1) = \sum_{i=1}^{3} u_i u_{i+1}^2 - \sum_{i=1}^{3} u_i^2 u_{i+1},$$
$$\Delta_2 = \sum(u_i^3 - 3u_i u_{i+1}^2) + 6u_1u_2u_3.$$

Since the form Δ_1 is only of degree 2 with respect to the quantities u_1, u_2, u_3, only the number 2 can occur as a common discriminant divisor. If we reduce the forms Δ_1 and Δ_2 modulo the divisor system

$$M = (2, u_1^2 - u_1, u_2^2 - u_2, u_3^2 - u_3),$$

we quickly see that Δ_1 contains it, and we get the following congruence for the primitive form $(\alpha\Delta_1 + \beta\Delta_2)$:

$$\alpha\Delta_1 + \beta\Delta_2 \equiv \beta[(u_1 + u_2 + u_3) + (u_1u_2 + u_2u_3 + u_3u_1)] \pmod{M},$$

which means the number 2 is only a common discriminant divisor when

$$\beta \equiv 0 \pmod{2}.$$

We can now give this result a more elegant form. The decomposition of ν in (6.) allows us to represent 4ν as:

$$4\nu = (2\alpha - 3\beta)^2 + 27\beta^2 = A^2 + 27B^2,$$

so that the number 4ν can always be expressed as $A^2 + 27B^2$, and the uniqueness of the decomposition in (6.) shows us that our depiction is also unique. From the two equations

$$A = 2\alpha - 3\beta, B = \beta$$

it follows that A and B are only divisible by 2 if the same is true for β, which means that in this case not only 4ν but also ν itself can be expressed in the form $A^2 + 27B^2$. So we have the following theorem:[127]

[126] Hensel doesn't say, but he knows and uses, that the three periods are cyclically permuted by the Galois action. The three forms w_1, w_2, w_3 below are, of course, Galois conjugates.

[127] Note that this is stated incorrectly in [**113**, 2.2.1, item 3].

If $\nu = 3\mu + 1$ is an arbitrary real prime, then the number 2 is a common inessential discriminant divisor in the genus $\mathfrak{G}_3(\epsilon)$, if and only if the number ν can be written in the form

$$\nu = A^2 + 27B^2.$$

For primes ν less than two hundred this is the case for cubic period equations which are formed by roots of unity of order[128]

$$31, 43, 109, 127, 157, 189.$$

§4

In his discussion mentioned above of the common inessential discriminant divisors of a genus, *Kronecker* calls attention to the remarkable circumstance that this can be eliminated,[129] if the coefficients $u_1, u_2, \dots u_n$ of the linear form

$$w_0 = u_1\xi_1 + \cdots + u_n\xi_n$$

are no longer in the domain of the real integers, but rather in the bigger realm of the algebraic numbers from another genus $\Gamma(\zeta)$.[130] *Kronecker* does not prove this interesting theorem, however, nor does he specify how the ring of integers should be chosen such that we can avoid the occurrence of a common discriminant divisor.[131] *Kronecker* did not return to this subject later, and so far I cannot find any hint of a proof in his papers.[132]

I would like to briefly touch on this point to show how to choose the genus $\Gamma(\zeta)$ for the coefficients $u_1, \dots, u_n$ so that the prime p is not a common discriminant divisor, and to determine the field of smallest degree for the adjunction. Finally, we wish to show that for this goal we only need the simplest algebraic numbers, which stem from the roots of unity of prime degree.

The following theorem, which is a simple extension of the one established in the previous sections, leads to these results:

---

[128] Sic, but of course $189 = 9^2 + 27 \cdot 2^2$ is not a prime. See [**115**, A014752] for primes of this form.

[129] Hensel is unclear on what exactly can be "eliminated" here, but he seems to mean the fact that the index form becomes divisible by p for all integer values.

[130] This observation from [**97**, §25] can be seen in the final paragraph of Section 5.4. Suppose p is a common inessential discriminant divisor for a number field. By the previous theorem, the index form is a primitive form in the variables u_i which becomes divisible by p whenever we replace each u_i by an integer $a_i \in \mathbb{Z}$. Kronecker's observation was that (in the specific example he was considering) we can obtain a value that is not divisible by p if we allow the values a_i to be algebraic integers in a larger field (which Hensel calls $\Gamma(\zeta)$). Hensel proposes to give a proof of Kronecker's remark and to explain how to find a field $\Gamma(\zeta)$.

[131] The discussion focuses entirely on the index form for K, without attempting to interpret it in terms of relative field extensions. Nevertheless, Hensel continues to use the language of "common discriminant divisors." We sketch a possible interpretation of Kronecker's "auxiliary field" in Section 6.5.2.

[132] This seems to indicate that Hensel waited until after Kronecker's death to publish these results because he expected to find such a proof in his papers.

Let

(1.) $$F_1(u_1),\ F_2(u_2),\ \ldots,\ F_n(u_n)$$

be n integral polynomials, each in one variable $u_1, \ldots, u_n$, such that each $F_i(u_i)$ modulo p has as many incongruent integer roots z_i as its degree. Further, let $F(u_1, \ldots, u_n)$ be an integral polynomial in all n indeterminates $u_1, \ldots, u_n$. Then the congruence

(2.) $$F(z_1, z_2, \ldots, z_n) \equiv 0 \pmod p$$

holds for all congruence roots $z_1, \ldots, z_n$ of the n functions (1.) if and only if $F(u_1, \ldots, u_n)$ contains the divisor system

$$(p; F_1(u_1), \ldots, F_n(u_n))$$

in the sense of *Kronecker*'s theory.

The proof of this theorem can be carried out in the same way as before, since the argument was based solely[133] on the fact that the congruences of degree p

$$F_i(u_i) = u_i^p - u_i \equiv 0 \pmod p$$

have exactly p incongruent roots modulo p, so as many as their degree, together with the fact that p is a prime number.

This theorem can finally also be extended[134] in the following way: we can assume that the coefficients of the $n+1$ functions $F_1(u_1), \ldots, F_n(u_n), F(u_1, \ldots, u_n)$ are no longer real integers, but rather elements of the ring of integers of a genus $\Gamma(\zeta)$ determined by an arbitrary algebraic integer ζ; only now we must work modulo a prime divisor $p(\zeta)$ of p instead of the element p, which inside $\Gamma(\zeta)$ loses the property of being indecomposable. Now if the functions[135]

$$F_i(u_i, \zeta)$$

modulo $p(\zeta)$ contain the same number of incongruent roots ζ_i inside the domain $\Gamma(\zeta)$ as their degree indicates, then we can show as before that for a polynomial $F(u_1, \ldots, u_n, \zeta)$, the congruences

$$F(\zeta_1, \zeta_2, \ldots, \zeta_n; \zeta) \equiv 0 \pmod{p(\zeta)}$$

will hold for all value systems $(\zeta_1, \ldots, \zeta_n)$ if and only if the polynomial $F(u_1, \ldots, u_n, \zeta)$ can be represented by the elements of the divisor system

$$(p(\zeta); F_1(u_1, \zeta), \ldots, F_n(u_n, \zeta))$$

in a homogeneous and linear way with integer coefficients.

---

[133]Working over $\mathbb{F}_p$, the argument boils down to the observation that modulo $F_1(u_1)$ the polynomial F becomes a polynomial of degree $\deg(F_1) - 1$ in u_1 whose coefficients are polynomials in the other variables. Replace $u_2, \ldots, u_n$ with arbitrarily chosen roots $z_2, \ldots, z_n$. We get a polynomial in u_1 which is zero for all possible choices of z_1. Since the number of choices is $\deg(F_1)$, so higher than the degree of the specialized polynomial, this polynomial must be identically 0. Hence each of the coefficient polynomials has the property that it is zero for all choices of $z_2, \ldots, z_n$. Now use induction.

[134]Now we allow the polynomials to have coefficients in some ring of integers and replace p by one of its prime divisors; this amounts to working over a finite extension of $\mathbb{F}_p$, and the argument goes through as before.

[135]Hensel does not just assume the coefficients are now in $\mathbb{Z}[\zeta]$, but rather indicates this explicitly in his notation.

We will now make a special assumption, that the *coefficients* of each of the $n+1$ function $F_i(u_i, \zeta)$ and $F(u_1, \dots, u_n, \zeta)$ are, as before, real integers.[136] To call attention to this assumption, we will denote them as before by $F_i(u_i)$ and $F(u_1, \dots, u_n)$. The congruence roots ζ_i of the function $F_i(u_i)$ modulo $p(\zeta)$, however, are still assumed to belong to the genus domain $\Gamma(\zeta)$. Now if we reduce the integral function $F(u_1, \dots, u_n)$ to the smallest remainder modulo the integral module system

$$(F_1(u_1), \dots, F_n(u_n)),$$

we get an integral function $\overline{F}(u_1, \dots, u_n)$ of $u_1, \dots, u_n$ with *real* integer coefficients, whose degree in u_i will always be smaller than the degree of the function $F_i(u_i)$. This reduced function can only be represented in a homogeneous way by the elements of the system

$$(p(\zeta), F_1(u_1), \dots, F_n(u_n)),$$

if all of their coefficients are divisible by $p(\zeta)$, that is to say by p itself. Thus $F(u_1, \dots, u_n)$ contains the divisor system

$$(p(\zeta), F_1(u_1), \dots, F_n(u_n))$$

if and only if it can be represented in a homogeneous and linear way by the real[137] system

$$(p, F_1(u_1), \dots, F_n(u_n))$$

The function $F(u_1, \dots, u_n)$ vanishes modulo $p(\zeta)$ for all congruence roots of the n functions $F_i(u_i)$ if and only if the congruence

$$F(u_1, \dots, u_n) \equiv 0 \quad [\text{modd. } (p, F_1(u_1), \dots, F_n(u_n))] \tag{3.}$$

is satisfied.

We can use this theorem to solve easily the question posed at the beginning of this paper. Take, as in the first paragraph of this paper,

$$w_0 = u_1\xi_1 + u_2\xi_2 + \cdots + u_n\xi_n$$

the fundamental form of the genus domain $(\mathfrak{G})$, and let $w_1, w_2, \dots, w_{n-1}$ denote the $n-1$ fundamental forms conjugate to w_0. Finally, let

$$\Delta^2(u_1, \dots, u_n)$$

be the discriminant of the fundamental equation freed from its numerical divisors (the genus discriminant[138]). Now if $\Gamma(\zeta)$ is another arbitrary genus domain and $p(\zeta)$ is a prime divisor of the real prime p in $\Gamma(\zeta)$, we can investigate under which conditions $p(\zeta)$ is a common inessential divisor of all discriminants $\prod(w_i - w_n)$, by now letting $u_1, \dots, u_n$ be arbitrary algebraic integers of the domain $\Gamma(\zeta)$ instead of arbitrary real integers. Equivalently, we can investigate the conditions under which the primitive form $\Delta(u_1, \dots, u_n)$ is always divisible by $p(\zeta)$, if we replace $u_1, \dots, u_n$ by arbitrary integers belonging to the domain $\Gamma(\zeta)$.

Now if k is the degree of the prime divisor $p(\zeta)$ for the genus Γ, then the number of integers[139] in Γ that are incongruent modulo $p(\zeta)$ is equal to p^k. So every number

[136] Hensel will assume the polynomials in question have rational integer coefficients, but still wants to allow the variables to take values in $\Gamma(\zeta)$. He claims that in this case the divisor $p(\zeta)$ above can in fact be replaced by p.

[137] Recall that Hensel uses "real" to mean "rational."

[138] As before Δ is the index form.

[139] Hensel always thinks in terms of representatives rather than congruence classes; so it's "the number of incongruent integers" rather than "the number of congruence classes."

ζ of this domain satisfies the congruence:

$$u^{p^k} - u \equiv 0 \pmod{p(\zeta)},$$

and so this congruence contains the same number of incongruent roots inside of Γ as its degree displays. The above question can now be stated as follows: Under what conditions is the primitive form $\Delta(u_1, \ldots, u_n)$ divisible by $p(\zeta)$ for the congruence roots modulo $p(\zeta)$ of the n functions

$$u_1^{p^k} - u_1,\ u_2^{p^k} - u_2,\ \ldots,\ u_n^{p^k} - u_n.$$

This question is directly answered by the last theorem, if we replace the n functions $F_i(u_i)$ by $u_i^{p^k} - u_i$. So we get the theorem:[140]

> The prime divisor $p(\zeta)$ over the domain of rationality $\Gamma(\zeta)$ is a common inessential divisor of all equation discriminants of $(\mathfrak{G})$ if and only if $\Delta(u_1, \ldots, u_n)$, the discriminant of $(\mathfrak{G})$ freed of its numerical factor, contains the divisor system
>
> $$P_k = (p; u_1^{p^k} - u_1, \ldots, u_n^{p^k} - u_n) \tag{4.}$$
>
> in *Kronecker*'s sense, where k is the degree of $p(\zeta)$ for the domain Γ.

Since the above criterion is solely dependent on the degree k of $p(\zeta)$, it applies equally to all divisors of p in Γ with the given degree.

It follows from this that over the rationality domain $\Gamma(\zeta)$ the prime divisor $p(\zeta)$ is not a common inessential discriminant divisor of $(\mathfrak{G})$ if and only if the following condition is satisfied:

$$\Delta(u_1, \ldots, u_n) \not\equiv 0 \pmod{(p; u_1^{p^k} - u_1, \ldots, u_n^{p^k})}.$$

This result can now be used to decide what assumptions need to be made on the values taken by coefficients $u_1, \ldots, u_n$ of the fundamental form of $(\mathfrak{G})$, $w_0 = u_1\xi_1 + \cdots + u_n\xi_n$, inside a domain $\Gamma(\zeta)$, so that the discriminant $D(w_0)$ of the n conjugate values $w_0, w_1, \ldots, w_{n-1}$ does not contain *any* prime factor of p other than those in the discriminant, or equivalently, whether it is possible to choose values for the unknowns $u_1, \ldots, u_n$ in $\Gamma(\zeta)$ so that the form $\Delta(u_1, \ldots, u_n)$ is relatively prime to p.

Clearly we will first need to choose $u_1, \ldots, u_n$, so that the primitive form $\Delta(u_1, \ldots u_n)$ is coprime to every prime divisor of p in $\Gamma(\zeta)$, i.e., so that none of the prime divisors [of p in $\Gamma(\zeta)$] are common inessential divisors of the discriminants $D(w_0)$ from $(\mathfrak{G})$. If

$$p = p_1^{e_1}(\zeta) \ldots p_l^{e_l}(\zeta)$$

is the decomposition of p into its prime factors in $\Gamma(\zeta)$, and if

$$k_1,\ \ldots,\ k_l$$

are the degrees of the individual distinct prime divisors, then p can only have the required properties if the primitive form $\Delta(u_1, \ldots, u_n)$ does not contain any of the l divisor systems

$$P_{k_i} = (p; u_1^{p^{k_i}} - u_1, \ldots, u_n^{p^{k_i}} - u_n) \qquad (i = 1, 2, \ldots, l),$$

[140] The statement is confusing because it refers to the "equation discriminants of $(\mathfrak{G})$" being divisible by a divisor in $\Gamma(\zeta)$. See Section 6.5.2 for our interpretation.

where of course we only need to investigate those systems for which the numbers k_i are distinct. If these conditions are satisfied, then it is easy to see that for the unknowns $u_1, \ldots, u_n$, such integers $\zeta_1, \ldots, \zeta_n$ of the genus domain $\Gamma(\zeta)$ can[141] be chosen so that the number $\Delta(\zeta_1, \ldots, \zeta_n)$ is co-prime to p. If each of the $p_i(\zeta)$ is not a common inessential divisor of the equation discriminant $D(w_0)$, then for each i we can find n numbers $\zeta_1^{(i)}, \ldots, \zeta_n^{(i)}$ such that

$$\Delta(\zeta_1^{(i)}, \ldots, \zeta_n^{(i)}) \not\equiv 0 \pmod{p_i(\zeta)}.$$

If we now consider these numbers for each of l prime divisors of p, we can choose other numbers $\zeta_1, \ldots, \zeta_n$ so that for each i we have

$$\zeta_1 \equiv \zeta_1^{(i)}, \zeta_2 \equiv \zeta_2^{(i)}, \ldots, \zeta_n \equiv \zeta_n^{(i)} \pmod{p_i(\zeta)} (i = 1, 2, \ldots, l),$$

and so for each i:

$$\Delta(\zeta_1, \ldots, \zeta_n) \equiv \Delta(\zeta_1^{(i)}, \ldots, \zeta_n^{(i)}) \not\equiv 0 \pmod{p_i(\zeta)}.$$

This means the number $\Delta(\zeta_1, \ldots, \zeta_n)$ is in fact coprime to p.[142]

We will call the genus $\Gamma(\zeta)$ a *supplementary genus* for the domain $\mathfrak{G}(\xi)$ with respect to the prime p if we can choose values in in $\Gamma(\zeta)$ for the unknowns $u_1, \ldots, u_n$ in the n conjugate fundamental forms $w_0, w_1, \ldots, w_{n-1}$ so that the discriminant $\prod(w_i - w_k)$ contains the prime p no more than the genus discriminant of $(\mathfrak{G})$. Then we can state the necessary and sufficient conditions for $\Gamma(\zeta)$ to be a supplementary genus for $\mathfrak{G}(\xi)$ with respect to p:

> Let $p = p_1^{\epsilon_1} p_2^{\epsilon_2}, \ldots, p_l^{\epsilon_l}$ be the decomposition of the real prime p into prime factors inside the domain Γ, and let $k_1, k_2, \ldots, k_\lambda$ be the distinct [residual] degrees [of the p_i]. Then $\Gamma(\zeta)$ is a supplementary genus for $(\mathfrak{G})$ with respect to the prime p if and only if the discriminant of the fundamental equation of $(\mathfrak{G})$ freed from its numerical factor does not contain any of the λ divisor systems:
>
> $$P_{k_1}, P_{k_2}, \ldots, P_{k_\lambda}.$$

If the adjoined genus $\Gamma(\zeta)$ is *Galois*, then the degrees of all prime divisors of p are equal. Setting k to be the common value of all the degrees, we can replace the more complicated condition above by the simpler condition that the form $\Delta(u_1, \ldots, u_n)$ does not contain a divisor system

$$P_k = (p; u_1^{p^k} - u_1, \ldots, u_n^{p^k} - u_n).$$

Further, I would also like to remark that in the above theorem all of the divisor systems P_k can be omitted if the index k is a multiple of one of the other numbers $k_1, \ldots, k_\lambda$. If

$$\Delta(u_1, \ldots, u_n) \not\equiv 0 \pmod{P_k},$$

then a fortiori

$$\Delta(u_1, \ldots, u_n) \not\equiv 0 \pmod{P_{ak}};$$

[141] Hensel claims here that if we know we can make each of a set of conditions hold separately, we can make them hold simultaneously. The "Chinese remainder theorem" argument is given in the rest of this paragraph.

[142] Added a paragraph break here.

because from the congruence

$$u^{p^{ak}} - u \equiv (u^{p^k} - u)^{p^{(a-1)k}} \equiv 0 \pmod{(p; u^{p^k} - u)}$$

it follows that every divisor system P_{ak} is a multiple of P_k: so $\Delta(u_1, \ldots, u_n)$ cannot be divisible by P_{ak} if it does not contain the system P_k.

§5

We should now discuss which is the supplementary genus $\Gamma(\zeta)$ of lowest degree for a given genus $(\mathfrak{G})$ with relation to an arbitrary prime p.

To this end, we investigate the discriminant of the fundamental equation freed from its numerical factor

$$\Delta(u_1, \ldots, u_n)$$

as to its divisibility by the divisor systems

$$P_1, P_2, P_3, \ldots,$$

where in each case

$$P_k = (p; u_1^{p^k} - u_1, \ldots, u_n^{p^k} - u_n).$$

If the primitive form does not already contain the first system

$$P_1 = (p; u_1^p - u_1, \ldots, u_n^p - u_n),$$

then p is not at all an inessential divisor for $(\mathfrak{G})$, which means the supplementary genus of lowest degree is that of natural integers. If however, $\Delta(u_1, \ldots, u_n)$ is divisible by P_1, in the sequence $P_1, P_2, \ldots$ we must come at last to a divisor system

$$P_k = (p; u_1^{p^k} - u_1, \ldots, u_n^{p^k} - u_n),$$

which is no longer contained[143] in Δ. Now choose μ large enough so that the power p^μ is bigger than $\frac{n(n-1)}{2}$, which is to say larger than the dimension[144] of the primitive form $\Delta(u_1, \ldots, u_n)$. Then this form cannot be reduced to one of lower degree modulo the system $(u_1^{p^\mu} - u_1, \ldots, u_n^{p^\mu} - u_n)$, because all exponents of $u_1, \ldots, u_n$ will be smaller than p^μ. Therefore, the form $\Delta(u_1, \ldots, u_n)$ could only contain the module system

$$(p; u_1^{p^\mu} - u_1, \ldots, u_n^{p^\mu} - u_n),$$

if all its coefficients were divisible by p, which contradicts the assumption that $\Delta(u_1, \ldots, u_n)$ is primitive. Thus, the form can only contain a finite number of divisor systems $P_1, P_2, \ldots$, and the number of systems it contains will be smaller than μ, where p^μ is the lowest power of p, which is bigger than $\frac{n(n-1)}{2}$.

Now take P_k to be the first module system of the series $P_1, P_2, \ldots, P_{k-1}, P_k$, which is not contained in Δ, so that each of the previous contain the form. Now if

$$\varphi(\zeta) = \zeta^k + a_1\zeta^{k-1} + \cdots + a_k = 0$$

[143] Hensel is stating a lemma: it is not possible that for all k the form Δ is a linear combination of the elements in P_k. The proof, given in the rest of this paragraph, is easy: no form can be a linear combination of polynomials of degree bigger than its own degree unless it is zero, so if $\Delta \in P_k$ for large enough k it would have to be divisible by p. But Δ is primitive. That puts an upper bound on k.

[144] As before, Hensel seems to use "dimension" for the degree of a homogeneous form.

is an equation of degree k, whose left side is also irreducible modulo p,* then the domain $\Gamma(\zeta)$ that it defines will be[145] a supplementary genus for $\mathfrak{G}(\xi)$ with respect to p, and indeed it will be one of the lowest possible degree.

Indeed, $\Gamma(\zeta)$ is a supplementary genus of $(\mathfrak{G})$. The function $\varphi(\zeta)$ is irreducible modulo p, so p is itself a prime inside the domain $\Gamma(\zeta)$ whose [residual] degree is[146] p^k. The prime p will be a common inessential divisor over the domain of rationality $\Gamma(\zeta)$ of the equation discriminants of $(\mathfrak{G})$ if and only if the form $\Delta(u_1,\ldots,u_n)$ contains the divisor system $P_k = (p; u_i^{p^k} - u_i)$, which contradicts the previous assumption.

Furthermore, if $\Gamma_1(\eta)$ is a different domain whose degree is smaller than k, and $p(\eta)$ is a prime divisor of p, then its degree k_1 is at most equal to the degree of $\Gamma_1(\eta)$, and so is smaller than k. Therefore $p(\eta)$ is a common inessential divisor for $(\mathfrak{G})$ in the domain $\Gamma_1(\eta)$, because the form $\Delta(u_1,\ldots,u_n)$ contains the divisor system $P_{k_1} = (p; u_i^{p^{k_1}} - u_i)$, whose index is smaller than k.[147] So we have the following theorem:

> If
> $$P_k = (p; u_1^{p^k} - u_1, \ldots, u_n^{p^k} - u_n)$$
> is the divisor system of lowest degree that is *not* contained in the primitive form $\Delta(u_1,\ldots,u_n)$, then the smallest supplementary genus Γ of $\mathfrak{G}(\xi)$ for the prime p has degree k. Such a domain will be defined by every polynomial equation of degree k, whose left side is irreducible modulo p.

From the proof it follows also that we can only obtain a supplementary genus defined by an equation of degree k if the left side is irreducible modulo p. Further,[148] the domain $\Gamma(\zeta)$ defined by the above equation, that is, the totality of rational functions of ζ, contains all functions defined by irreducible equations of degree k modulo p, and so we see that there is only one domain $\Gamma(\zeta)$, which is a supplementary genus domain of lowest degree with respect to $\mathfrak{G}(\xi)$.

This last statement should be understood as follows. The algebraic integers in two genera of degree k [defined by equations] which are irreducible modulo p, are pairwise congruent modulo this prime, so that for the question we are considering

[145] This claim is proved in the next two paragraphs. Notice that Hensel doesn't care if the supplementary field is linearly disjoint from his original field; that only matters if we want to interpret his result in terms of a relative extension KL/L; see Section 6.5.2. So he can use any equation of degree k which is irreducible mod p.

[146] Sic, but Hensel means k. He says "ihre Ordnung für denselben ist p^k," literally "the order for the same is p^k," so perhaps he means the number of congruence classes?

[147] Since all he really needs is for the residual degree of at least one of the factors of p to be equal to k, the smallest possible degree for the supplementary field is realized when p is inert and $f = k$.

[148] Hensel seems to be claiming that there is only one field $\Gamma(\zeta)$ of this type, but that is not true. The residue fields will all be $\mathbb{F}_{p^k}$, of course, and that may be what he is referring to. See the next paragraph, where he admits he doesn't really mean what he has said.

*A polynomial of degree k that is irreducible modulo p always exists because the number

$$\bar{g}(k) = \frac{1}{k}\sum_{d|k} \epsilon_\delta p^d$$

of such functions is never equal to zero.

one domain can be substituted for the other.[149] The algebraic character of the supplementary genera Γ can, however, be very different, which raises the question of which is the algebraically simplest[150] supplementary genus for a given domain $(\mathfrak{G})$ with respect to the prime p.

In the previously discussed place[151] in his Festschrift, *Kronecker* considers[152] the genus defined by the cubic equation

$$\alpha^3 - \alpha^2 - 2\alpha - 8 = 0,$$

for which the number 2 is a common inessential discriminant divisor (as was first noted by Mr. *Dedekind* in the cited paper[153]). [Kronecker suggests] that this prime stops being an inessential divisor if the domain Γ of the third roots of unity is adjoined to the rational domain.[154] For this domain $(\mathfrak{G})$, then, the very simple *cyclotomic* genus of the third roots of unity is a supplementary genus with respect to the inessential divisor 2.

This suggests an interesting theorem, that for every domain $(\mathfrak{G})$ and an arbitrary prime p we can find a supplementary genus of the greatest algebraic simplicity, namely one constructed from[155] roots of unity of prime degree. In what follows we prove this theorem and develop a method to find the smallest such domain.[156]

Let $(\mathfrak{G})$ be a domain of nth degree and let $\Delta(u_1, \ldots, u_n)$ be the discriminant of the fundamental equation of $(\mathfrak{G})$ freed of its numerical factor.[157] Further let

$$P_{k_1}, P_{k_2}, \ldots P_{k_l}$$

be the divisor systems

$$P_k = (p; u_1^{p^k} - u_1, \ldots, u_n^{p^k} - u_n)$$

in which the primitive form $\Delta(u_1, \ldots, u_n)$ contains.[158] We need only consider those whose index k is not contained in one of the other indexes $k_1, \ldots, k_i$ as a divisor, because according to the observation above, the form $\Delta(u_1, \ldots, u_n)$ (if contained

[149] So what Hensel means is that if we have two fields L_1 and L_2 defined by irreducible polynomials that remain irreducible modulo p, then p remains prime in both L_1 and L_2, and in both cases the residue field is $\mathbb{F}_{p^k}$. This makes the two fields equivalent "for the question we are considering." In modern terms, L_1 and L_2 have the same p-adic completion.

[150] Hensel doesn't say what he means by "algebraically simplest," of course. It will turn out that he can always choose a cyclotomic field.

[151] See the final paragraph of Section 5.4.

[152] Hensel says something like "Kronecker suggests that for this field... for which 2... (as was first noted...), that... That was too much, so we broke it up.

[153] This is **[23]**; see Chapter 4.

[154] With the integral basis $\{1, \alpha, 4/\alpha = \frac{1}{2}(\alpha^2 + \alpha) - 1\}$ given by Dedekind, the index form in this case is $\Delta(u_1, u_2, u_3) = 2u_2^3 - u_2^2 u_3 - u_2 u_3^2 - 2u_3^3$, which is clearly always even for integer values of the u_i. If ζ is a cube root of unity, however, $\Delta(1, \zeta, \zeta^2) = 1$.

[155] It seems likely that when Hensel says "constructed from roots of unity" he means a subfield of a cyclotomic field. But one could also ask for a cyclotomic field rather than a subfield. In fact, in what follows Hensel first finds the smallest prime-order cyclotomic field that has the desired property, then finds the smallest subfield of that field that still has the property.

[156] To clarify the argument, we exemplify using the cubic field above. We use the integral basis $\{1, \alpha, \beta\}$ where $\beta = 4/\alpha$.

[157] The discriminant of the fundamental equation is $-503(2u_2^3 - u_2^2 u_3 - u_2 u_3^2 - 2u_3^3)^2$. While Hensel never says it explicitly, to find his Δ we first divide by the field discriminant (here, -503) and then take the square root, so in this case $\Delta(u_1, u_2, u_3) = 2u_2^3 - u_2^2 u_3 - u_2 u_3^2 - 2u_3^3$, as mentioned above. That 2 is a common inessential discriminant divisor follows from $\Delta(u_1, u_2, u_3) = -u_3(u_2^2 - u_2) - u_2(u_3^2 - u_3) + 2(u_2^3 + u_3^3 - u_2 u_3)$.

[158] In our example there is only one, with $k = 1$.

in the system P_k) is also divisible by every system P_d, whose index is a divisor of k_i. So we form the whole number:

$$F(p) = (p^{k_1} - 1)(p^{k_2} - 1) \dots (p^{k_l} - 1)$$

and look for the smallest prime ν different from p which is not contained in $F(p)$.[159] Then the domain Γ_ν of the ν-th roots of unity is the smallest which is a supplementary genus for $(\mathfrak{G})$ with respect to the prime p.[160]

Now we easily prove that Γ_ν is in fact a supplementary genus for $(\mathfrak{G})$. If p modulo ν belongs to the exponent k, then p decomposes in Γ_ν into distinct prime factors of degree k. So Γ_ν is a supplementary genus of $(\mathfrak{G})$ if and only if $\Delta(u_1, \dots, u_n)$ does not contain the divisor system P_k. This claim is only satisfied if the index k is not a divisor of any of the l numbers $k_1, k_2, \dots, k_l$. If this were however the case, then at least one of the l factors of the product $F(p)$, and hence also $F(p)$ itself, would be divisible by ν. Since ν is the smallest prime not occurring in the product, the first part of the claim[161] is proved.

Further, if ν_1 is a prime smaller than ν and if Γ_{ν_1} is the genus constituted by the ν_1th roots of unity, then Γ_{ν_1} can not be a supplementary genus of $(\mathfrak{G})$. In fact, according to the previous assumption, ν_1 is a divisor of the product $F(p)$, which means ν_1 is contained in at least one of the factors $(p^{k_i} - 1)$. Therefore, the exponent k' of p modulo ν_1 is a divisor of one of the l numbers $k_1, \dots, k_l$, and so the divisor system P_k is a divisor of the form $\Delta(u_1, \dots, u_n)$, or equivalently, p is an inessential divisor for $(\mathfrak{G})$ in the domain Γ_{ν_1}.[162]

Now[163] take Γ_ν to be the supplementary genus of roots of unity we have just determined, of lowest degree. Then it will contain as many genus domains $\Gamma_\nu(\lambda)$ as the number of divisors of $\nu - 1$. Namely, if

$$\nu - 1 = \lambda\mu$$

is a decomposition of $\nu - 1$ in two factors, then under ν there is contained a domain of degree λ, namely the one containing the λ periods of μ terms formed from the ν-th roots of unity. We should now investigate which period genus[164] $\Gamma_\nu(\lambda)$ is of lowest degree λ and is still a supplementary genus.[165]

To answer this question,[166] we think of all the divisors of $\nu - 1$ arranged according to their value in descending order and denote them by

$$\mu_1, \mu_2, \mu_3, \dots, \mu_\varrho,$$

so that $\mu_1 = \nu - 1$, $\mu_\varrho = 1$, and

$$\mu_1 > \mu_2 > \mu_3 > \dots > \mu_\varrho.$$

---

[159] In our example $F(2) = (2^1 - 1) = 1$, so $\nu = 3$.

[160] This result is to be proved in the next few paragraphs. Note that in our example it is, as Kronecker pointed out, the field of cube roots of unity.

[161] So Γ_ν is a supplementary genus.

[162] So no cyclotomic field corresponding to a smaller prime ν_1 will do the job.

[163] We have found a cyclotomic field that will serve as a supplementary genus. Now Hensel wants to show that we can take a specific subfield.

[164] Periodengattung.

[165] So $\Gamma_\nu(\lambda)$ is the unique cyclic subfield of degree λ in the cyclotomic field of ν-th roots of unity. The "periods" are those defined by Gauss in the last chapter of his *Disquisitiones Arithmeticae*; they give an explicit basis of the field $\Gamma_\nu(\lambda)$.

[166] Hensel first states the result: we choose the largest divisor of $\nu - 1$ that satisfies a divisibility condition.

Substituting p in $F(p)$ by each element of the sequence $p^{\mu_1}, p^{\mu_2}, \ldots, p^{\mu_\varrho}$, we see first that $F(p^{\mu_1})$ is divisible by ν, because every one of the factors

$$p^{\mu_1 k_l} - 1 = (p^{r-1})^{k_l} - 1$$

contains this prime. But $F(p^{\mu_\varrho}) = F(p)$ does not contain this prime, according to the above assumption on ν. Let then μ be the first, and therefore largest, of these numbers such that

$$F(p^{\mu}) = (p^{\mu k_1} - 1) \ldots (p^{\mu k_l} - 1)$$

is no longer divisible by ν. Let λ be the complementary divisor of $\nu - 1$ to μ, that is,

$$\lambda\mu = \nu - 1.$$

Then the genus $\Gamma_\nu(\lambda)$ containing the λ periods with μ terms of the νth roots of unity, is the smallest[167] which is still a supplementary genus for $(\mathfrak{G})$.

That this is a supplementary genus, one sees as follows: If κ is the exponent of p^{μ} modulo ν, that is, the smallest whole number for which the difference $(p^{\kappa\mu} - 1)$ is divisible by ν, then p decomposes in $\Gamma_\nu(\lambda)$ into distinct prime factors of degree κ. The domain $\Gamma_\nu(\lambda)$ is then a supplementary genus of $(\mathfrak{G})$ if the primitive form $\Delta(u_1, \ldots, u_n)$ is not contained the module system P_κ, so if κ is not among the numbers $k_1, \ldots, k_l$. But if this were the case, then one of the factors $p^{\mu k_i} - 1$, and so the number $F(p^{\mu})$, would be divisible by ν, which is not the case.[168]

Now taking a different period domain $\Gamma_\nu(\lambda')$ of smaller degree, so that $\lambda' < \lambda$, and since $\lambda'\mu' = \nu - 1$, so that $\mu' > \mu$, then this cannot be a supplementary genus for $(\mathfrak{G})$. Namely, if κ' is the exponent of $p^{\mu'}$ modulo ν, then $p^{\mu'\kappa'} - 1$ is divisible by ν, and κ' must be a divisor of one of the numbers $k_1, \ldots, k_l$. Since $F(p^{\mu'})$ is divisible by ν for every $\mu' > \mu$, one of the factors $(p^{\mu k_i} - 1)$ is a multiple of ν, which means the exponent κ' belonging to $p^{\mu'}$ modulo ν is contained in k_i, and from this it follows that p is still an inessential divisor of the discriminants of $(\mathfrak{G})$ over the domain of rationality $\Gamma_\nu(\lambda')$. With this, we have proved the above conjecture, and we can summarize the end result of all the last investigations in the following elegant theorem:

[167] This is the claim. The proof follows. In our example, of course, the only divisors of $3 - 1 = 2$ are $\mu_1 = 2$ and $\mu_2 = 1$ and the only subfield that works is Γ_3 itself.

[168] So we have shown that the chosen $\Gamma_\nu(\lambda)$ is a supplementary genus. It remains to show that it is the smallest.

Let $(\mathfrak{G})$ be a given genus of degree n and let $\Delta(u_1, \dots, u_n)$ be the discriminant of the fundamental equation freed from its numerical factor. Further let p be any real prime and denote by

$$P_{k_1}, P_{k_2}, \dots, P_{k_l}$$

the divisor systems of the form

$$P_k = (p; u_1^{p^k} - u_1, \dots, u_n^{p^k} - u_n)$$

which Δ contains, chosen so that among their indexes $k_1, \dots, k_l$ none is a multiple of the others.
Let ν be the smallest prime which does not divide the integer

$$F(p) = (p^{k_1} - 1) \dots (p^{k_l} - l).$$

Then the genus Γ_ν of the ν-th roots of unity is the smallest which is a supplementary genus of the genus $(\mathfrak{G})$. Further, if μ is the largest divisor of $\nu - 1$, for which the number

$$F(p^\mu)$$

does not contain the prime ν, the period genus $\Gamma_\nu(\lambda)$ contained in it, defined by the λ periods of μ terms in the νth roots of unity, is the smallest which still has this property.

Berlin, November 17, 1893

6.5. Hensel's Achievements

Hensel's paper establishes two criteria for a prime to be a common inessential discriminant divisor. The first is the same as Dedekind's 1878 result: a prime p is a CIDD if and only if one cannot find sufficiently many distinct irreducible polynomials in $\mathbb{F}_p[x]$ to reflect the factorization of p. The second is new to Hensel: a prime p is a CIDD if the index form has values that are always divisible by p.

In both cases, Hensel goes beyond the statement of a criterion, however. He also provides a way to test each of his criteria. In §1 he emphasizes that the number of irreducible polynomials of a given degree can be computed explicitly. In §2 he explains how one can test whether p satisfies Dedekind's criterion *without* knowing the factorization of (p). In §3 he determines exactly when a primitive polynomial in n variables has values that are always divisible by p. The result is explicitly computable, so that once again we have a way to test whether p is a CIDD.

Both of the criteria confirm that CIDDs are a "small primes" phenomenon. Hensel's result in §3 implies at once that if p is a CIDD in a field of degree n, then $p \leq \frac{1}{2}n(n-1)$. As Żyliński showed in [**145**], Dedekind's criterion gives the stronger estimate $p < n$.

While Hensel would later revisit his results (for example in [**80**]) in terms of his new theory of p-adic numbers, this article mostly concludes the story. The existence of common inessential discriminant divisors had moved from being an obstacle to directly extending Kummer's methods to general number fields to being an expected property of small primes.

6.5.1. Hensel's Construction in §2.

Perhaps the most interesting aspect of Hensel's paper is §2, where he constructs ideals (or divisors) that allow us to determine, for any prime p, how many prime ideals of degree f appear in the factorization of (p). In this section we outline the construction, giving examples that highlight some aspects of its properties.

Suppose we have a number field K of degree n over $\mathbb{Q}$. We write $\mathscr{O}$ for its ring of integers. We choose an integral basis

$$\{1, \xi_2, \xi_3, \ldots, \xi_n\}.$$

Everything we will do depends on the choice of integral basis.

Let p be a rational prime. We want to determine, for each f, how many distinct prime ideals of degree f appear in the factorization of (p) in $\mathscr{O}$. We now set $\xi_1 = 1 + p$.

Choose an integer k between 1 and n, and let I_k be the ideal generated by $(\xi_i^{p^k} - \xi_i)$, $1 \leq i \leq n$. We are interested in the primes above p that also divide the ideal I_k. The argument is entirely local at p, that is, we completely ignore the rest of the factorization of I_k. What Hensel shows is that each prime ideal $\mathfrak{p}$ that divides p and has residual degree dividing k will divide I_k exactly once.

From there, it is a matter of using Möbius inversion. Define the fractional ideals

$$F_f = \prod_{k|f} I_k^{\mu(f/k)},$$

where μ is the Möbius function. Then F_f is divisible exactly once by each prime ideal of degree f that divides p. If λ_f is the number of those prime ideals, then the p-part of the norm of F_f has valuation $f\lambda_f$. Thus, one can determine λ_f by constructing the ideals F_f, computing their norms, and determining their valuations.

The ideals I_k and F_f depend on the choice of integral basis, but Hensel's theorems show that their p-parts do not. The p-part of the ideal F_f is integral, but Hensel does not consider the question of whether F_f is always an integral ideal.

Let us consider first the "most dramatic example" mentioned in footnote 88. Suppose ℓ is a prime number, $p \equiv 1 \pmod{\ell}$, and K is the ℓ-th cyclotomic field. It follows from Kummer's theory that p is the product of $\ell - 1$ primes of degree one. What does Hensel's construction yield?

Let ζ be an ℓ-th root of unity, and take the integral basis

$$\{1, \zeta, \zeta^2, \ldots, \zeta^{\ell-2}\}.$$

We have $\xi_1 = 1 + p$ and $\xi_i = \zeta^{i-1}$ for $i = 2, 3, \ldots, \ell - 1$. Since $p \equiv 1 \pmod{\ell}$, we have $\xi_i^{p^k} - \xi = 0$ when $i \geq 2$. So I_k is simply the ideal generated by $(1+p)^{p^k} - (1+p)$. This integer is always divisible by p but not by p^2, and so the p-part of the norm of $F_1 = I_1$ is $p^{\ell-1}$ and all the other F_f are prime to p.

Suppose, however, we switch to the other standard integral basis

$$\{\zeta, \zeta^2, \ldots, \zeta^{\ell-1}\}.$$

(This disobeys Hensel's instruction to always choose an integral basis that includes 1, but the results will still hold.) Now we set $\xi_1 = \zeta + p$ and we will have to consider the ideals J_k generated by $(\zeta + p)^{p^k} - (\zeta + p)$. The p-parts of I_k and J_k will be the same, but the two ideals will not be equal.

To take a simple example, let $\ell = 3$ and $p = 7$. We use SageMath to do the computation in $\mathbb{Q}(\zeta)$, where $\zeta^3 = 1$. Note first that

$$(7) = (-\zeta - 3)(\zeta - 2).$$

I_1 is the ideal generated by $8^7 - 8 = 2097144$, which factors as

$$2^3(\zeta - 8)(-\zeta - 3)(\zeta - 2)(\zeta + 2)^6(-2\zeta + 3)(2\zeta + 5)(\zeta + 9),$$

while J_1 is the ideal generated by $(\zeta + 7)^7 - (\zeta + 7)$, which factors as

$$2^2(\zeta)(\zeta - 3)(-\zeta - 3)(\zeta - 2)(\zeta + 2)^4(\zeta + 4)(-2\zeta + 3)(\zeta + 6)(\zeta + 7).$$

Each, as expected, is divisible by 7 exactly once, but the other factors are completely different.

For a more interesting example, let K be the number field obtained by adjoining a root α of the polynomial $x^4 + x^3 + 6x^2 + 2x + 12$. (Global field 4.0.13564.1 in [**109**].) An integral basis is then

$$1, \xi_2 = \alpha, \xi_3 = \frac{1}{2}(\alpha^2 + \alpha^3), \xi_4 = \alpha^3.$$

The discriminant of K is $13564 = 2^2 \cdot 3391$; the class number is 1, so all ideals are principal.

Let $p = 2$ and set $\xi_1 = 1 + p = 3$. We want to consider the ideals I_k generated by $\xi_i^{2^k} - \xi_i$ for $k = 1, 2, 3, 4$. Computing with SageMath we see that

$$I_1 = I_2 = I_3 = (2\alpha^3 + 3\alpha^2 + 3\alpha + 12)$$

and

$$I_4 = (\alpha^2 - \alpha).$$

Hensel showed, in §2 of his paper (page 107), that a prime p is ramified if and only if none of the ideals I_k is divisible by p. That is the case here: neither one of our ideals is divisible by 2, so it follows that 2 is ramified in K.

The ideals Hensel calls F_κ are then

$$F_1 = I_1, \quad F_2 = I_2 I_1^{-1} = (1), \quad F_3 = I_3 I_1^{-1} = (1),$$

and

$$F_4 = I_4 I_2^{-1} = (\alpha^3 - \alpha^2 - 11).$$

Their norms are

$$N(F_1) = 24, \qquad N(F_2) = N(F_3) = 1, \qquad N(F_4) = 11.$$

Notice that this shows that the ideals F_κ do have divisors that are not divisors of $p = 2$.

Hensel showed that the 2-adic valuation of the norm of F_κ is equal to $\kappa\lambda_\kappa$, where λ_κ is the number of divisors of 2 that have degree κ. So we get $\lambda_1 = 3$ and $\lambda_2 = \lambda_3 = \lambda_4 = 0$, i.e., 2 is divisible by three distinct ideals of degree one. Since there are only two irreducible polynomials of degree one in $\mathbb{F}_2[x]$, it follows that 2 is a common inessential discriminant divisor in K. Of course, since 2 is ramified in K, it is *also* an essential divisor of the discriminant.

In this case, we can conclude the form of the factorization of 2 in the field K: we must have $(2) = \mathfrak{p}_1^2\mathfrak{p}_2\mathfrak{p}_3$ with all prime factors of degree 1. What we cannot find from Hensel's construction are the three ideals $\mathfrak{p}_i$, nor, of course, which of them appears to degree two in the factorization.

If we switch from $p = 2$ to $p = 3$, the results are

$$N(F_1) = 96 = 2^5 \cdot 3, \qquad N(F_2) = 5, \qquad N(F_3) = 27 = 3^3, \qquad N(F_4) = 55,$$

showing that $\lambda_1 = 1$, $\lambda_2 = 0$, $\lambda_3 = 1$, and $\lambda_4 = 0$. So (3) is the product of two prime ideals, one of degree one and the other of degree three. The ideal I_3 is divisible by 3, but I_1, I_2, and I_4 are not, which confirms that 3 is not ramified in K.

It is worth pointing out that even with modern technology the explicit computation of the I_κ quickly becomes infeasible. With $p = 3$, $\kappa = 4$, for example, we are already looking at ideals generated by terms of the form $\xi_i^{81} - \xi_i$. This is a commonly-noted aspect of Kronecker's (and Hensel's) constructionist philosophy: it is important that it be *theoretically* possible to compute things, but there is no requirement that the construction be efficient.

6.5.2. Hensel's "supplementary fields". In the last two sections of his paper, Hensel addresses a conjecture of Kronecker about the values of the index form. If a prime number p is a CIDD, the index form $\Delta(u_1, u_2, \dots, u_n)$ will be a primitive homogeneous polynomial which nevertheless has the property that for every choice of n-tuple $(a_1, a_2, \dots, a_n) \in \mathbb{Z}^n$ the value $\Delta(a_1, a_2, \dots, a_n)$ is divisible by p. Kronecker noticed, in the example he was discussing, that if we replace $\mathbb{Z}$ with the ring of integers of some larger number field, then we can get values that are *not* divisible by p.

This is in fact easy to see given Hensel's results in §3, which allow him to determine when a primitive polynomial in several variables has the property that all of its values are divisible by p. He shows two things: first, such behavior follows from the congruence $a^p \equiv a \pmod{p}$, which is true for all $a \in \mathbb{Z}$. In terms of the theory of finite fields, the point is that the residue field $\mathbb{F}_p$ is fixed by the Frobenius automorphism. Second, Hensel shows in §3 that there are no multivariable effects: one need only look at $u_i^p - u_i$ for each i. One of the first results in §4 is that this remains true when we plug in integers from a larger field, except that we must replace the exponent p by a power of p determined by the size of the residue field.

If we want to remove the property, then, all that we need to do is to extend the base field to a field L such that the residue field(s) above p are bigger. (How much bigger is determined by the highest power of p such that the polynomial belongs to the ideal generated by $u_i^{p^k} - u_i$.) Such an L he calls a supplementary genus to the field (genus) K. In §4 and §5, Hensel shows that a supplemetary genus always exists, showing that in fact he can choose L to be a subfield of a cyclotomic field unramified at p.

Hensel talks about this in terms of CIDDs: "this prime stops being an inessential divisor if the domain Γ of the third roots of unity is adjoined to the rational domain," for example. This suggests he is saying something about a relative extension: instead of the field $K/\mathbb{Q}$ for which p is a common index divisor, consider the extension KL/L.

It seems unlikely, however, that at this point Hensel was in possession of a good theory of relative extensions; all of his papers discuss number fields as finite extensions of $\mathbb{Q}$. There is no sign that he realized that there might be an issue if K and L are too closely related. (At the extreme, we might even have $K = L$.) He simply seems to have assumed that the index form for the field K over $\mathbb{Q}$ is also the index form for KL over L, whatever that might mean.

Since, as Hensel showed, we have a lot of freedom in choosing the field L, we can give an interpretation in terms of CIDDs after placing some restriction on the choice of L.

Denote the original number field, in which p is a CIDD, by K. Assume that K and L are *linearly disjoint*. Then the integral basis $\{\xi_1, \xi_2, \ldots, \xi_n\}$ of K over $\mathbb{Q}$ will also be an integral basis for KL over L and the relative discriminant $d(KL/L)$ is the ideal in $\mathcal{O}_L$ generated by d_K. In this situation, Hensel's result means that p is no longer a common index divisor for KL/L. That is, there exists an element $\theta \in KL$ such that the index of $\mathcal{O}_L[\theta]$ in $\mathcal{O}_{KL}$ is not divisible by p.

CHAPTER 7

The Rock in the Middle of the Road

After Kummer had created a theory of ideal prime divisors for algebraic integers in cyclotomic fields, most mathematicians saw no great need to extend it to general number fields. Those who did give the matter some thought seemed to feel that it would be a fairly easy matter, at least for most primes: the defining polynomial of the field provides congruential conditions that can be used to define the ideal prime divisors needed. It was known (even to Kummer) that this would not work for primes dividing the discriminant of the defining polynomial. Kummer (in his theory of "Kummer extensions") and others (for example, Bachmann on biquadratic fields) seemed willing to leave it that way.

Kronecker and Dedekind both tell us that they set out to create a full theory of divisors along Kummer's path. Perhaps they both realized that the problem did not affect all the primes dividing the polynomial discriminant, but only those that divided the index. (This was first stated by Dedekind in 1871; see Chapter 3 above.) Even after that insight, however, the obstacle remained: in some cases, no matter which defining polynomial one chose, there would be primes for which the congruential approach failed: these were the "common inessential discriminant divisors."

Brazilian poet Drummond de Andrade wrote a poem[1] that captures well the emotional effect of running into such an obstacle. "In the Middle of the Road" the narrator finds, there is a rock.[2] The narrator's shock is captured by the repetition in many forms of the main experience: the way is blocked, in the middle of the road there is a rock. Dedekind and Kronecker both seem to remember their discovery of CIDDs in a similar way.

There was a choice. One could attempt to use Kummer's approach for all but the evil primes, and then create a special theory to handle those cases. This is what Zolotarev attempted to do. Both Dedekind and Kronecker decided, however, that the technical difficulty posed by CIDDs indicated that the congruential approach was the wrong one. As Kronecker said, it did not reflect "the nature of things." Both set out to find a path that allowed all primes to be treated the same way. They left the blocked road entirely and found another.

As is well known, each came up with a different path, influenced by their different views of how mathematics should be done and also by their distinct ultimate goals: Dedekind seemed mostly interested in generalizing Kummer to other number fields, while Kronecker wanted a unified theory of all "algebraic quantities." Dedekind went for set-theoretic constructions while Kronecker preferred actual formulas,

[1]"No Meio do Caminho" in **[40]**. There have been many English translations.

[2]The Portuguese word is "pedra," which can mean "rock" or "stone"; the translators have chosen variously.

polynomials in several variables. The differences between the two approaches have been extensively discussed in the historical literature.

"Does it really need to be so complicated?" seems to have been the question both needed to answer.[3] As we saw above, both pointed to CIDDs as the answer to that question; their existence blocked the natural path. As we saw in Chapter 4, when Dedekind became aware of Zolotarev's first paper he felt the need to point out that there was no way around the problem.

To demonstrate the power of their new approaches, it remained to clarify the issue of CIDDs. When and why did they occur? Were they a rare exception or common? Dedekind and Hensel[4] both did this in the papers we have presented.

As a result, CIDDs became much less interesting: they turned out to simply reflect the fact that some primes are small. It was soon shown that if a prime p was a CIDD in a field of degree n, then $p < n$. Further, it was shown that for any prime p and any integer $n > p$ there exist fields of degree n for which p is a CIDD.

All cyclotomic fields and all quadratic fields are monogenic, i.e., there exist an elements θ such that $\mathbb{Z}[\theta]$ is the full ring of integers. In that case, of course, there are no CIDDs. Much of the early intuition was built upon these examples, which perhaps explains the initial surprise about CIDDs, which yielded the first examples of non-monogenic fields. There are, however, many non-monogenic fields for which CIDDs do not exist. This seems, in fact, to be the most common situation; see, for example, [**1**]. One should note, however, that this paper also suggests that the existence of CIDDs is much less common than the failure of monogenicity. As a result, modern work focuses more on the property of being monogenic than on the existence of CIDDs.

Of course, no mathematical subject ever dies out completely. Nowadays one defines the *index* $i(K)$ of a number field K to be the greatest common divisor of all the indices $(\mathscr{O} : \mathbb{Z}[\theta])$ as θ runs through the algebraic integers in K. A prime is a CIDD if and only if it divides $i(K)$. In the online database of number fields at [**109**], the index is one of the invariants listed. See [**113**, pp. 74–77] for a few more results about the index. One can, of course, ask further questions. Is it true that for n large enough most fields of degree n have $i(K) > 1$?

The issue also remains significant in computational algebraic number theory. For primes not dividing the index, Dedekind's theorem gives a simple algorithm for determining factorizations. For other primes, however, one needs something else. See the discussion in [**16**, Sections 4.8 and 6.2].

In the early period we have focused on, number fields were almost always considered as extensions of $\mathbb{Q}$. In the years that followed, the relative situation, where there is a base field F and a finite extension K/F, became important. The questions discussed in this book can certainly be asked and answered in the relative setting as well, but we have avoided following that path.

In Drummond's poem, the rock is remembered: the narrator says he will never forget the event. Our story, however, ends here: from being the rock in the middle

[3]According to [**11**, p. 397], Hilbert and Hurwitz found the methods of both Dedekind and Kronecker "scheußlich" (awful).

[4]It seems that Kronecker's focus remained on algebraic geometry, where CIDDs never occur. Kronecker had proved this for function fields in one variable (in characteristic zero, of course) in [**100**]. His definition of the field discriminant in the case of dimension bigger than one excluded the possibility of CIDDs. This is probably why he left it to Hensel to push forward on the algebraic number theory side of the problem.

of the road in the 1850s, common inessential discriminant divisors became the explanation for the different approaches of Dedekind and Kronecker in the 1870s and 1880s. The new theories allowed Dedekind and Hensel to characterize and conquer them, and the rock was largely forgotten.

Bibliography

[1] Levent Alpöge, Manjul Bhargava, and Ari Shnidman. A positive proportion of quartic fields are not monogenic yet have no local obstruction to being so. *Mathematische Annalen*, 388:4037–4052, 2024.

[2] Jeremy Avigad. Methodology and metaphysics in the development of Dedekind's theory of ideals. In Ferreirós and Gray **[52]**, pages 159–186.

[3] Raymond Ayoub. The lemniscate and Fagnano's contributions to elliptic integrals. *Archive for History of Exact Sciences*, 29:131–149, 1984.

[4] Paul Bachmann. *Die Theorie der complexen Zahlen welche aus zwei Quadratwurzeln zusammengesetzt sind.* W. Weber & Co., 1867.

[5] Paul Bachmann. Zur Theorie der complexen Zahlen. *Journal für die Reine und Angewandte Mathematik*, 67:200–204, 1867.

[6] Richard Balzer. *Theorie und Anwendung der Determinanten.* Hirzel, fourth edition, 1875.

[7] I. G. Bashmakova and A. N. Rudakov. Algebra and Algebraic Number Theory. In Kolmogorov and Ushkevich **[96]**, chapter 2.

[8] M. Bauer. Über den ausserwesentlicher Discriminantenteiler algebraischer Körper. *Mathematische Annalen*, pages 573–576, 64.

[9] Oswald Baumgart. *The Quadratic Reciprocity Law: A Collection of Classical Proofs.* Edited and translated by Franz Lemmermeyer. Birkhäuser, 2015.

[10] Kurt-R. Biermann. *Die Mathematik und ihre Dozenten an der Berliner Universität, 1810–1930.* Akademie-Verlag, second edition, 1988. The first edition (1973) covered 1810–1920.

[11] Otto Blumenthal. Lebensgeschichte. In *David Hilbert Gesammelte Abhandlungen* **[88]**, pages 388–429.

[12] Z. I. Borevich and I. R. Shafarevich. *Number Theory.* Academic Press, 1966.

[13] Maria Teresa Borgato, Erwin Neuenschwander, and Irène Passeron, editors. *Mathematical Correspondences and Critical Editions.* Birkhäuser, 2018.

[14] Augustin-Louis Cauchy. Mémoire sur les fonctions qui ne peuvent obtenir que deus valeurs. *Journal de l'École Polytechnique*, 1815. Read to the Institute, November 20, 1812. Reprinted in *Oeuvres Complètes d'Augustin Cauchy*, 2ème série, Tome 1.

[15] Karine Chemla. Fragments of a history of the concept of ideal. `arXiv:2307.16234`, July 2023.

[16] Henri Cohen. *A Course in Computational Algebraic Number Theory.* Springer, 1996.

[17] David A. Cox. *Primes of the Form $x^2 + ny^2$: Fermat, Class Field Theory, and Complex Multiplication.* AMS Chelsea, third edition, 2022.

[18] Richard Dedekind. Abriß einer Theorie der höheren Kongruenzen in bezug auf einem reelen Primzahl-Modulus. *Journal für die Reine und Angewandte Mathematik*, 54:1–26, 1857. In **[25]**, Vol. 1, item V.

[19] Richard Dedekind. Anzeige der erste Auflage von Dirichlets Vorlesungen über Zahlentheorie. *Göttingische gelehrte Anzeigen*, pages 121–124, 1864. In **[25]**, Vol. 3, item LIII.

[20] Richard Dedekind. Anzeige der zweiten Auflage von Dirichlets Vorlesungen über Zahlentheorie. *Göttingische gelehrte Anzeigen*, pages 1481–1494, 1871. In **[25]**, Vol. 3, item LV.

[21] Richard Dedekind. Supplement X: Ueber die Composition der binären quadratische Formen. In *Vorlesungen über Zahlentheorie* **[35]**. Edited and with supplements by Richard Dedekind.

[22] Richard Dedekind. Sur la théorie des nombres entiers algébriques. *Bulletin des Sciences Astronomiques et Mathématiques*, 1876–1877. Partially reprinted in **[25]**, Vol. 3, item XLVIII. English translation **[26]**.

[23] Richard Dedekind. Über den Zusammenhang zwischen der Theorie der Ideale und der Theorie der höheren Kongruenzen. *Abhandlungen der Königlichen Gesellschaft der Wissenschaften zu Göttingen*, 23, 1878. (Each paper has its own page numbering; Dedekind's is the third paper in the mathematical section of the volume.) In **[25]**, Vol. 1, item XV.

[24] Richard Dedekind. Über die Diskriminanten endlicher Körper. *Abhandlungen der Königlichen Gesellschaft der Wissenschaften zu Göttingen*, 29, 1882. (Each paper has its own page numbering; Dedekind's is the second paper in the mathematical section of this volume.) In **[25]**, Vol. 1, item XIX.

[25] Richard Dedekind. *Gesammelte Mathematische Werke*. Chelsea, 1969. Ed. by R. Fricke, E. Noether, und O. Ore. Originally Published in Braunschweig, 1930–32.

[26] Richard Dedekind. *Theory of Algebraic Integers*. Translation of **[22]** with an extended introduction by John Stillwell. Cambridge University Press, 1996.

[27] Richard Dedekind and Heinrich Weber. Theorie der algebraichen Functionen einer Veränderlichen. *Journal für die Reine und Angewandte Mathematik*, 92:181–290, 1882. See the translations **[28]** and **[29]**.

[28] Richard Dedekind and Heinrich Weber. *Theory of Algebraic Functions of One Variable*. Translated and introduced by John Stillwell. American Mathematical Society, 2012.

[29] Richard Dedekind and Heinrich Weber. *Théorie des Fonctions Algèbriques d'une Variable*. Translated, introduced, and annotated by Emmylou Haffner. J. Vrin, 2019.

[30] B. N. Delone. *The St. Petersburg School of Number Theory*. American Mathematical Society, 2005.

[31] Jean Dieudonné, editor. *Abrégé d'Histoire des Mathématiques*. Hermann, 1978.

[32] P. G. Lejeune Dirichlet. Recherches sur les formes quadratiques à coefficients et à indeterminées complexes. *Journal für die Reine und Angewandte Mathematik*, 24:291–371, 1842. Reprinted in **[37]**, Vol. I, pp. 533–618.

[33] P. G. Lejeune Dirichlet. De formarum binariarum secundi gradus compositione. *Journal für die Reine und Angewandte Mathematik*, 47:155–160, 1851. Reprinted in **[37]**, Vol. II, pp. 105–114.

[34] P. G. Lejeune Dirichlet. *Vorlesungen über Zahlentheorie*. Edited and with supplements by Richard Dedekind. Vieweg, Braunschweig, 1st edition, 1863.

[35] P. G. Lejeune Dirichlet. *Vorlesungen über Zahlentheorie*. Edited and with supplements by Richard Dedekind. Vieweg, Braunschweig, 2nd edition, 1871.

[36] P. G. Lejeune Dirichlet. Einige Resultate von Untersuchungen über einer Classe homogener Functionen des dritten und der höheren Grade. In *G. Lejeune Dirichlet's Werke* **[37]**. Volume I, Item XXXIV.

[37] P. G. Lejeune Dirichlet. *G. Lejeune Dirichlet's Werke*. Edited by L. Kronecker, two volumes. Druck und Verlag von Georg Reimer, Berlin, 1889, 1897. Reprinted in one volume by Chelsea, 1969.

[38] P. G. Lejeune Dirichlet. Zur Theorie der complexen Einheiten. In *G. Lejeune Dirichlet's Werke* **[37]**. Volume I, Item XXXVI.

[39] P. G. Lejeune Dirichlet. *Lectures on Number Theory*. Translated by John Stillwell. American Mathematical Society, 1999. Main text and first nine supplements of **[35]**.

[40] Carlos Drummond de Andrade. *Alguma Poesia*. Compania das Letras, 2013. Originally published 1930.

[41] H. M. Edwards. Kronecker's arithmetical theory of algebraic quantities. *Jahresbericht der Deutschen Mathematiker-Vereinigung*, 94(3):130–139, 1992.

[42] Harold M. Edwards. The background of Kummer's proof of Fermat's last theorem for regular primes. *Archive for History of Exact Sciences*, 14(3):219–236, 1975.

[43] Harold M. Edwards. The genesis of ideal theory. *Archive for the History of the Exact Sciences*, 23:321–378, 1980.

[44] Harold M. Edwards. Dedekind's invention of ideals. *The Bulletin of the London Mathematical Society*, 15(1):8–17, 1983. Reprinted as **[45]**.

[45] Harold M. Edwards. Dedekind's invention of ideals. In Esther R. Phillips, editor, *Studies in the History of Mathematics*. Mathematical Association of America, 1987.

[46] Harold M. Edwards. *Divisor Theory*. Birkhäuser, Boston, 1990.

[47] Harold M. Edwards. Mathematical ideas, ideals, and ideology. *The Mathematical Intelligencer*, 14(2):6–19, 1992.

[48] Harold M. Edwards. *Fermat's Last Theorem: A Genetic Introduction to Algebraic Number Theory*, volume 50 of *Graduate Texts in Mathematics*. Springer-Verlag, New York, 1996.

[49] H. T. Engstrom. The theorem of Dedekind in the ideal theory of Zolotarev. *Transactions of the American Mathematical Society*, 32:879–887, 1930.

[50] Moritz Epple. *Die Entstehung der Knotentheorie: Kontexte und Konstruktionen einer modernen mathematischen Theorie*. Vieweg, 1999.

[51] C. C. Gillispie et al., editor. *Complete Dictionary of Scientific Biography*. Scribners, 2008. Electronic resource.

[52] José Ferreirós and Jeremy J. Gray, editors. *The Architecture of Modern Mathematics*. Oxford, 2006.

[53] Daniel E. Flath. *Introduction to Number Theory*. AMS Chelsea Publishing, 2018. With errata. Reprint of the Wiley (1989) edition.

[54] Günther Frei. The unpublished section eight: On the way to function fields over a finite field. In Goldstein et al. **[66]**, chapter II.4.

[55] Otto Frostman. Aus dem Briefwechsel von G. Mittag Leffler. In H. Behnke and K. Kopfermann, editors, *Festschrift zur Gedächnisfeier für Karl Weierstraß, 1815–1965*, pages 53–56. Westdeutscher Verlag, 1966.

[56] Évariste Galois. Sur la théorie des nombres. *Bulletin des Sciences Mathématiques, Physiques et Chimiques*, 13:428–435, 1830. Reprinted in *Journal de Mathématiques Pures et Apliquées*, 1846, 398–407. Item II.4 in **[114]**.

[57] Carl Friedrich Gauss. *Werke*, volume I. Königlichen Gesselschaft der Wissenchaften zu Göttingen, 1863. Reprinted in the Cambridge Library Collection by Cambridge University Press.

[58] Carl Friedrich Gauss. *Werke*, volume III. Königlichen Gesselschaft der Wissenchaften zu Göttingen, 1866. Reprinted in the Cambridge Library Collection by Cambridge University Press.

[59] Carl Friedrich Gauss. *Disquisitiones Arithmeticae*. Springer, 1986. Translated by Arthur A. Clarke from Gauss's *Werke*, volume 1, second edition (1870).

[60] Carolo Frederico Gauss. Demonstratio nova altera theorematis omnem functionem algebraicam rationalem integram unius variabilis in factores reales primi vel secundi gradus resolvi posse. *Comentationes Societatis Regiae Scientificarum Gottingensis Recentiores*, III, 1825. Reprinted in **[58]**.

[61] Catherine Goldstein. Johann Peter Gustav Lejeune-Dirichlet, *Vorlesungen über Zahlentheorie*, First Edition (1863). In I. Grattan-Guinness, editor, *Landmark Writings in Western Mathematics 1640–1940*, pages 480–490. Elsevier, 2005.

[62] Catherine Goldstein. The Hermitian form of reading the *Disquisitiones*. In Goldstein et al. **[66]**.

[63] Catherine Goldstein. Hermite and Lipschitz: A correspondence and its echoes. In Borgato et al. **[13]**.

[64] Catherine Goldstein and Norbert Schappacher. A book in search of a discipline (1801–1860). In Goldstein et al. **[66]**, chapter I.1.

[65] Catherine Goldstein and Norbert Schappacher. Several disciplines and a book (1860–1901). In Goldstein et al. **[66]**, chapter I.2.

[66] Catherine Goldstein, Norbert Schappacher, and Joachim Schwermer, editors. *The Shaping of Arithmetic after C. F. Gauss's Disquisitiones Arithmeticae*. Springer, 2007.

[67] Fernando Q. Gouvêa. Learning to move with Dedekind. In Dick Jardine and Amy Shell-Gellasch, editors, *Mathematical Time Capsules*. Mathematical Association of America, 2011.

[68] Emmylou Haffner. The "science of numbers" in action in Richard Dedekind's works. Dissertation, Université Paris Diderot–Paris 7, 2014.

[69] Emmylou Haffner. Strategical use(s) of arithmetic in Richard Dedekind and Heinrich Weber's *Theorie der algebraischen Funktionen einer Veränderlichen*. *Historia Mathematica* 44:31–69, 2017.

[70] Helmut Hasse. Existenztheoreme über algebraische Zahlkörper. *Mathematische Annalen*, 20:267–279, 1925.

[71] Helmut Hasse. Kurt Hensel zum Gedächtnis. *Journal für die Reine und Angewandte Mathematik*, 187:1–13, 1950.

[72] Helmut Hasse. *Number Theory*. Springer, 1978.

[73] Ralf Haubrich. Zur Entstehung der algebraischen Zahlentheorie Richard Dedekinds. Dissertation, Göttingen, 1982.

[74] Kurt Hensel. Arithmetische Untersuchungen über Discriminanten und ihre ausserwesentlichen Teiler. Dissertation, University of Berlin, 34-page booklet, 1884.

[75] Kurt Hensel. Untersuchung der ganzen algebraischen Zahlen eines Gattungsbereiches für einen beliebigen algebraischen Primdivisor. *Journal für die Reine und Angewandte Mathematik*, 101:99–141, 1887.

[76] Kurt Hensel. Über die Darstellung der Zahlen eines Gattungsbereiches für einen beliebigen Primdivisor. *Journal für die Reine und Angewandte Mathematik*, 103:230–237, 1888.

[77] Kurt Hensel. Über die Darstellung der Determinante eines Systems, welches aus zwei andern componirt ist. *Acta Mathematica*, 14:317–319, 1891.

[78] Kurt Hensel. Arithmetische Untersuchungen über die gemeinsamen ausserwesentlichen Discriminantenteiler einer Gattung. *Journal für die Reine und Angewandte Mathematik*, 113:128–160, 1894.

[79] Kurt Hensel. Untersuchung der Fundamentalgleichung einer Gattung für eine reelle Primzahl als Modul und Bestimmung der Teiler ihrer Discriminante. *Journal für die Reine und Angewandte Mathematik*, 113:61–83, 1894.

[80] Kurt Hensel. Über die Fundamentalgleichung und die ausserwesentlichen Discriminantenteiler eines algebraischen Körpers. *Nachrichten von der Gesellschaft der Wissenschaften zu Göttingen. Mathematisch-Physikalische Klasse*, 1897:254–260, 1897.

[81] Kurt Hensel. Über die arithmetischen Eigenschaften der Faktoriellen. *Archiv der Mathematik und Physik. 3. Reihe*, 2:293–294, 1902.

[82] Kurt Hensel. *Theorie der algebraischen Zahlen I*. Teubner, 1908.

[83] Charles Hermite. Sur la théorie des formes quadratiques: Premier mémoire. *Journal für die Reine und Angewandte Mathematik*, 47:313–342, 1854. Reprinted in *Oeuvres de Charles Hermite*, Vol. 1.

[84] Charles Hermite. Extrait dâĂŹune lettre de M. C. Hermite à M. Borchardt sur le nombre limité dâĂŹirrationalités auxquelles se réduisent les racines des équations à coefficients entiers complexes dâĂŹun degré et dâĂŹun discriminant donnés. *Journal für die Reine und Angewandte Mathematik*, 53:182–192, 1857. Reprinted in *Oeuvres de Charles Hermite*, Vol. 1.

[85] Charles Hermite. *Oeuvres*, volume I. Edited by Émile Picard. Gauthier-Villars, 1905.

[86] Charles Hermite. Sur l'introduction des variables continues dans la théorie des nombres. In *Oeuvres* **[85]**, pages 164–192. Edited by Émile Picard.

[87] David Hilbert. Die Theorie der algebraischen Zahlkörper. *Jahresbericht der Deutschen Mathematiker-Vereinigung*, 1887.

[88] David Hilbert. *Gesammelte Abhandlungen, Vol. 3*. Springer, 1935.

[89] David Hilbert. *The Theory of Algebraic Number Fields*. Translation by Iain T. Adamson of **[87]**. Springer, 1998.

[90] Christian Houzel. Fonctions elliptiques et intégrales Abéliennes. In Dieudonné **[31]**, pages 1–113. Reprinted in **[91]**.

[91] Christian Houzel. *La Géométrie Algébrique: Recherches Historiques*. Albert Blanchard, 2002.

[92] Christian Houzel. The work of Niels Henrik Abel. In *The legacy of Niels Henrik Abel. Papers from the Abel bicentennial conference, University of Oslo, Oslo, Norway, June 3–8, 2002*, pages 21–177. Springer, 2004.

[93] Christian Houzel. Elliptic functions and arithmetic. In Goldstein et al. **[66]**.

[94] Carl Gustav Jacob Jacobi. Ueber die complexen Primzahlen, welche in der Theorie der Reste der 5ten, 8ten und 12ten Potenzen zu betrachten sind. *Journal für die Reine und Angewandte Mathematik*, 19:314–318, 1839.

[95] Carl Gustav Jacob Jacobi. *Vorlesungen über Zahlentheorie—Wintersemester 1836/37, Königsberg*, volume 62 of *Algorismus. Studien zur Geschichte der Mathematik und der Naturwissenschaften*. With a foreword by Franz Lemmermeyer and Herbert Pieper. Dr. Erwin Rauner Verlag, Augsburg, 2007.

[96] A. N. Kolmogorov and A. P. Ushkevich, editors. *Mathematics of the 19th Century: Mathematical Logic, Algebra, Number Theory, Probability Theory*. Birkhäuser, 1992.

[97] Leopold Kronecker. Grundzüge einer arithmetischen Theorie der algebraischen Grössen. *Journal für die Reine und Angewandte Mathematik*, 92:1–123, 1882. Originally published by Reimer, Berlin, 1882. Reprinted in **[98]**, Vol. 2, item XI. Page numbers refer to the latter printing.

[98] Leopold Kronecker. *Mathematische Werke*. Edited by Kurt Hensel. Leipzig, 1895–1930. Edited by Kurt Hensel. Reprinted by AMS Chelsea Publishing, 1968.

[99] Leopold Kronecker. Mémoire sur les facteurs irreductibles de l'expression $x^n - 1$. In *Mathematische Werke* **[98]**, pages 75–92. Originally published in *Journal de Mathématiques Pures et Apliquées*, Series I, **19** (1854), p. 177–192.

[100] Leopold Kronecker. Ueber die Discriminante algebraischer Functionen einer Variabeln. In *Werke* **[98]**, pages 193–236. Read to the Berlin Academy in 1862; published in *Journal für die Reine und Angewandte Mathematik*, **91** (1881), 301–334.

[101] Leopold Kronecker. *Vorlesungen über Zahlentheorie*, volume 1. Edited by Kurt Hensel. Springer, 1978. Originally published by Teubner, Leipzig, 1901.

[102] E. E. Kummer. *Collected Papers*, volume I: Contributions to Number Theory. Springer, 1975.

[103] E. E. Kummer. Über die allgemeinen Reciprocitätgesetze unter den Resten und Nichresten der Potenzen, deren Grad eine Primzahl ist. In *Collected Papers I* **[102]**. Item 52. Original publication 1859.

[104] E. E. Kummer. Zur theorie der complexen Zahlen. In *Collected Papers I* **[102]**. Item 21. Original publication 1847.

[105] Adrien-Marie Legendre. *Essai su la Théorie des Nombres, Seconde Édition*. Courcier, Paris, 1808. Reprinted in the Cambridge Library Collection by Cambridge University Press.

[106] Franz Lemmermeyer. *Reciprocity Laws from Euler to Eisenstein*. Springer, 2000.

[107] Franz Lemmermeyer. Jacobi and Kummer's ideal numbers. *Abhandlungen aus dem Mathematischen Seminar der Universität Hamburg*, 79:165–187, 2009.

[108] Rudolf Lipschitz. *Briefwechsel mit Cantor, Dedekind, Helmholtz, Kronecker, Weierstrass*. Edited by Winfried Scharlau. Dokumente zur Geschichte der Mathematik, Band 2. Deutsche Mathematiker-Vereinigung, 1986.

[109] LMFDB Collaboration. The L-functions and modular forms database. `www.lmfdb.org`.

[110] Barry Mazur. Review of **[102]**. *Bulletin of the American Mathematical Society*, 83:976–988, 1977.

[111] Uta C. Merzbach. *Dirichlet: A Mathematical Biography*. Birkhäuser, 2018.

[112] Thomas Muir. *The Theory of Determinants in the Historical Order of Development*. Dover Publications, 1960. Four volumes bound as two. Originally published 1906, 1911, 1920, 1923.

[113] Władisław Narkiewicz. *The Story of Algebraic Numbers in the First Half of the 20th Century: From Hilbert to Tate*. Springer, 2018.

[114] Peter M. Neumann. *The Mathematical Writings of Évariste Galois*. European Mathematical Society, 2011.

[115] OEIS Foundation. Online Encyclopedia of Integer Sequences. `www.oeis.org`.

[116] Erwan Penchèvre. *Histoire d'Élimination Algébrique*. Garnier, 2021.

[117] Birgit Petri. *Perioden, Elementarteiler, Transzendenz — Kurt Hensels Weg zu den p-adischen Zahlen*. PhD thesis, Technischen Universität Darmstadt, 2011.

[118] Paola Piazza. Egor Ivanovitch Zolotarev and the theory of ideal numbers for algebraic number fields. *Rendiconti del Circolo Matematico di Palermo Serie II Supplemento*, 61:123–150, 1999.

[119] Paola Piazza. Zolotarev's Theory of Algebraic Numbers. In Goldstein et al. **[66]**, chapter VII.2.

[120] Adrian Rice. In search of the "birthday" of elliptic functions. *The Mathematical Intelligencer*, 30(2):48–56, 2008.

[121] David E. Rowe. *A Richer Picture of Mathematics: The Göttingen Tradition and Beyond*. Springer, 2018.

[122] Ranjan Roy. *Elliptic and Modular Functions: From Gauss to Dedekind to Hecke*. Cambridge, 2017.

[123] SageMath Developers. *SageMath, the Sage Mathematics Software System (Versions 9.2–10.2)*, 2010–2024. `www.sagemath.org`.

[124] Norbert Schappacher. On the history of Hilbert's twelfth problem: a comedy of errors. In *Matériaux pour l'histoire des mathématiques au XXe siècle (Nice, 1996)*, volume 3 of *Sémin. Congr.*, pages 243–273. Soc. Math. France, Paris, 1998.

[125] T. Schönemann. Grundzüge einer allgemeinen Theorie der höheren Congruenzen, deren Modul eine reelle Primzahl ist. *Journal für die Reine und Angewandte Mathematik*, 31:269–325, 1846.

[126] T. Schönemann. Von denjenigen Moduln, welche Potenzen von Primzahlen sind. *Journal für die Reine und Angewandte Mathematik*, 32:93–105, 1846.

[127] Eduard Selling. Ueber die idealen Primfactoren der complexen Zahlen, welche aus den Wurzeln einer beliebigen irreductiblen Gleichung rational gebildet werden. *Zeitschrift für Mathematik und Physik*, 10:17–47, 1865.

[128] J. A. Serret. *Cours d'Algèbre Supérieure.* Mallet-Bachelier, Paris, 1854.

[129] H. J. S. Smith. *Report on the Theory of Numbers.* Chelsea, 1965(?). Originally published in six parts as a Report of the British Association.

[130] J. J. Sylvester. On a remarkable discovery in the theory of canonical forms and of hyperdeterminants. *Philosophical Magazine*, pages 391–410, 1851. Reprinted in **[131]**, volume I, item 41.

[131] J. J. Sylvester. *The Collected Mathematical Papers of James Joseph Sylvester.* Chelsea, 1973. Original publication 1904–1912.

[132] N. Tchebotarev. The foundations of the ideal theory of Zolotarev. *American Mathematical Monthly*, 37:117–128, 1930.

[133] P. Ullrich. Die Entdeckung der Analogie zwischen Zahl- und Funktionenkörpern: der Ursprung der "Dedekind-Ringe". *Jahresbericht der Deutschen Mathematiker-Vereinigung*, 101:116–134, 1999.

[134] Peter Ullrich. On the origins of p-adic analysis. In *Proceedings of the 2nd Gauss Symposium. Conference A: Mathematics and Theoretical Physics (Munich, 1993)*, Sympos. Gaussiana, pages 459–473. de Gruyter, Berlin, 1995.

[135] Peter Ullrich. The genesis of Hensel's p-adic numbers. In *Charlemagne and his heritage. 1200 years of civilization and science in Europe, Vol. 2 (Aachen, 1995)*, pages 163–178. Brepols, Turnhout, 1998.

[136] Peter Ullrich. Karl Weierstraß and the theory of Abelian and elliptic functions. In *Karl Weierstraß (1815–1897)*, pages 143–163. Springer, 2016.

[137] Cédric Vergnerie. L'Algèbre sans les "fictions des racines": Kronecker et la théorie des caractéristiques dans les *Vorlesungen über die Algebraischen Gleichungen*. *Révue d'Histoire des Mathématiques*, 25:1–107, 2019.

[138] S. G. Vlăduţ. *Kronecker's Jugendtraum and modular functions.* Translated from the Russian by M. Tsfasman. Gordon and Breach, 1991.

[139] André Weil. La cyclotomie jadis et naguère. In *Œuvres Scientifiques/Collected Papers* **[141]**, pages 311–326. Corrected second printing, 1980.

[140] André Weil. Number theory and algebraic goometry. In *Œuvres Scientifiques/Collected Papers* **[141]**, pages 442–452. Corrected second printing, 1980.

[141] André Weil. *Œuvres Scientifiques/Collected Papers.* Springer, 1979. Corrected second printing, 1980.

[142] E. I. Zolotarev. *The Theory of Integral Complex Numbers with an Application to the Integral Calculus.* PhD thesis, St. Petersburg, 1874. Abstracted in *Fortschritte der Mathematik*, **6** (1874), p. 177.

[143] E. I. Zolotarev. Sur les nombres complexes. *Bulletin de l'Académie Impériale des Sciences Saint-Petersbourg*, (3) 24: columns 310–317, 1878.

[144] E. I. Zolotarev. Sur la théorie des nombres complexes. *Journal de Mathématiques Pures et Appliquées*, 6:51–84, 129–166, 1880.

[145] E. von Źyliński. Theorie der ausserwesentlichen Diskriminantenteiler algebraischer Körper. *Mathematische Annalen*, 73:273–274, 1913.

Index

PUBLISHED TITLES IN THIS SUBSERIES

47 **Fernando Q. Gouvêa and Jonathan Webster,** Common Inessential Discriminant Divisors, 2025

39 **Richard Dedekind and Heinrich Weber,** Theory of Algebraic Functions of One Variable, 2012

37 **Henri Poincaré,** Papers on Topology, 2010

30 **Michael Rosen, Editor,** Exposition by Emil Artin: A Selection, 2006

29 **J. L. Berggren and R. S. D. Thomas,** Euclid's *Phaenomena*, 2006

25 **J. M. Plotkin, Editor,** Hausdorff on Ordered Sets, 2005

19 **Hermann Grassmann,** Extension Theory, 2000

17 **Jacques Hadamard,** Non-Euclidean Geometry in the Theory of Automorphic Functions, 2000

16 **P. G. L. Dirichlet and R. Dedekind,** Lectures on Number Theory, 1999

Zeitfracht Medien GmbH
Ferdinand-Jühlke-Straße 7
99095 Erfurt, Deutschland
produktsicherheit@kolibri360.de